WJEC Physics
for A2 Level
Revision Workbook

Gareth Kelly

Iestyn Morris

Nigel Wood

Published in 2022 by Illuminate Publishing Limited, an imprint of Hodder Education, an Hachette UK Company, Carmelite House, 50 Victoria Embankment, London EC4Y 0DZ

Orders: please contact Hachette UK Distribution, Hely Hutchinson Centre, Milton Road, Didcot, Oxfordshire, OX11 7HH. Telephone: +44 (0)1235 827827. Email: education@hachette.co.uk. Lines are open from 9 a.m. to 5 p.m., Monday to Friday. You can also order through our website: www.hoddereducation.co.uk

British Library Cataloguing in Publication Data

A catalogue record for this book is available from the British Library

ISBN 978-1-912820-64-1

Printed by Ashford Colour Press Ltd

Impression 2
Year 2024

Hachette UK's policy is to use papers that are natural, renewable and recyclable products and made from wood grown in well-managed forests and other controlled sources. The logging and manufacturing processes are expected to conform to the environment regulations of the country of origin.

Publisher: Eve Thould

Editor: Geoff Tuttle

Design and layout: Nigel Harriss

Cover image: © Shutterstock / Pavel L Photo and Video

Acknowledgements

The authors would like to thank Adrian Moss for guiding us through the early stages of production and the Illuminate team of Eve Thould, Geoff Tuttle and Nigel Harriss for their patience and meticulous attention to detail. We are also indebted to Dr Sarah Ryan for her eagle eyes in spotting mistakes and inconsistencies and for her many insightful suggestions.

Picture Credits

p148, Prof. J. Leveille / Science Photo Library

All other illustrations © Illuminate Publishing

Contents

How to use this book

What this book contains

The A2 physics course, being the second year of the A level physics course, consists of two theory units and a practical unit.

Unit 3: Oscillations and Nuclei – divided into six topics – assessed by a series of questions in Section A of an exam paper, worth 80 marks. Section B consists of a comprehension passage on a physics topic followed by a series of questions based on the comprehension passage for 20 marks. The comprehension passage is usually two pages long and will involve some new physics related to some part of the course. Section B is meant to be synoptic so the questions will mainly relate to Units 1, 2 and 4 rather than Unit 3. The total time of the examination is 2 hours 15 minutes.

Unit 4: Fields and Options – divided into five compulsory topics and four options – assessed by a 2-hour exam. The compulsory topics form Section A of the exam and have an allocation of 80 marks. Section B is for the options and is worth 20 marks – you will study one of the options.

Unit 5: Practical examination – consisting of an experimental task and a practical analysis task.

This book has practice questions for Units 3 and 4, which includes questions on practical skills including the specified practical work.

Each topic in Units 3 and 4 has its own section in this book and there you will find:

- A concept map, which displays how the different concepts within the topic are related to one another and to other topics.
- A set of graded questions, with space for your answers, which are designed to test the content of the topic in a way which is similar to the examination.
- A question and answers section containing one or two examples of exam-style questions, with answers from two students, Rhodri and Ffion (who produce answers of different standard), together with examiner marks and discussion.

The next section comprises two practice papers – one for each unit. The final section consists of model answers to all the graded questions and the practice papers.

How to use this book effectively

This book can be used exclusively for revision, in which case you will work your way gradually through the questions as you revise each topic. Alternatively, it can be used regularly as you are working your way through A level physics and each topic of questions can be used as an end of topic check/test. Your teacher might even like to use it as a homework book because this book, potentially, contains 15 sets of end of topic homework questions as well as the practice papers.

Probably the worst method of revising physics is simply reading your notes or your textbooks, a technique that will simply send you to sleep and will not help you retain more than a small fraction of the information. Regardless of the quality of writing of the notes/textbooks, there is an inherent, soporific effect to reading notes. One way of combating the tendency to lose concentration is to make your own notes as you read through the notes/textbook. In simply reading notes, you will just prepare for exam questions in which you reproduce learned material. These, so-called Assessment Objective 1 (AO1) questions form only 30% of the exams. The higher-level skills required for AO2 and AO3 require different revision techniques. See pages 5 to 7 for further information.

You will find that the best way of revising is to answer exam-type questions. As in many fields, it is practice that makes perfect in physics and you should practise as many past papers and questions within this book as you can.

You will, inevitably, come across questions that seem difficult. Should you not be able to answer these questions, it is time to visit your notes/textbook but you will be doing so with a purpose to help you to concentrate. Should the question still leave you stumped, have a look at the answers in the back of the book. If the answer seems unintelligible, then it is time to ask your teacher or your fellow students for an explanation of how and why the answer is correct. Your teacher may also point out that different answers are creditworthy, especially in discussion questions.

So answering the questions within these covers and analysing the model answers will do more for your exam preparation than just reading notes or making revision timetables (however colourful)!

Assessment objectives

You need to demonstrate expertise in answering examination questions in different ways. One of the ways in which an exam question can be looked at is whether it is mathematical or not – being physics, there is maths in most questions. Some questions require expertise in practical skills, even in the written papers. However, besides these two categories are the assessment objectives. There are three of them.

Assessment Objective 1 (AO1)

AO1 questions are ones in which you need to:

demonstrate knowledge and understanding of scientific ideas, processes, techniques and procedures

These questions account for 28% of the marks in the two A2 unit papers. In the full A level, the tally is 30% because the AS units consist of less of the other assessment objectives.

The sentence in bold sounds far more complicated and highfalutin than is necessary. Essentially, these are the marks that can be obtained without too much thinking. This category covers:

● Recall of definitions, laws and explanations from the specification

● Inserting appropriate data into equations

● Deriving equations where required in the specification

● Describing experiments from the **specified practicals**.

In general, no judgement or new thinking is required. One can learn a definition or a law without fully understanding it, so it is a good idea to use the WJEC terms and definitions (T&D) booklet to help you memorise these. Defined T&D from the booklet are printed in **bold** in the concept map for each topic in this book.

Examples of AO1 questions

1. Explain why light nuclei tend to undergo fusion reactions rather than fission.   [4]

Good answer (4/4): Binding energy per nucleon (BE/N) is a very good measure of nuclear stability. Light nuclei have low BE/N and are therefore unstable. The most stable nuclei are around iron-56 or nickel-62. When lighter nuclei fuse, they will be increasing the nucleon number and increasing their BE/N. This means that the fusion products are more stable and energy is released in the process. If a light nucleus underwent fission, it would be moving away from stability and would be massively endothermic.

Bad answer (1/4): The bigger the nucleus the more stable it is, so fusion of nuclei always results in a nucleus that is more stable. Except for when you have things like uranium that prefers to do fission for releasing energy.

The *bad answer* is incorrect in stating that the larger the nucleus, the more stable it is because nuclei become less stable after a nucleon number of around 60. The student seems to understand that fusion should lead to more stable nuclei and might understand the link between increasing stability and energy release but one mark is all that this response merits.

2. State what is meant by simple harmonic motion.   [2]

Good answer (2/2): It is an oscillating motion that results from an acceleration which is proportional to displacement from a fixed point and directed towards it.

Bad answer (0/2): It's when the acceleration is proportional to distance.

The bad answer misses the idea that the distance is from a fixed point and mentions nothing about the direction of the acceleration.

Assessment Objective 2 (AO2)

AO2 questions are ones in which you need to:

apply knowledge and understanding of scientific ideas, processes, techniques and procedures:
1. **in a theoretical context**
2. **in a practical context**
3. **when handling qualitative data**
4. **when handling quantitative data.**

Here, the key words are 'apply knowledge'. The application of knowledge is required in theoretical, practical, qualitative and quantitative contexts. Theoretical just means some idealised context made up by the examiner. Practical means that the data have apparently come from a real experiment (although the data are usually made up by the examiner). Qualitative means without numbers and calculations whereas quantitative means the opposite (i.e. with numbers and calculations).

Note that application of knowledge here can also include analysis of data even though 'analyse' appears in AO3 (see page 8). Generally, if you are told what type of analysis to carry out, these will be AO2 skills. If the question is more open-ended and you must choose the analysis methods yourself, the question will be classified as AO3. This is the most common type of question and accounts for 44% of the marks on the papers. Note that all calculations must be mainly AO2 marks: we have seen that inserting data into an equation is classed as AO1, but any manipulation of an equation, such as changing the subject, and the production of a final answer is AO2.

Examples of AO2 questions

1. A parallel plate capacitor has plates that are square metal sheets separated by 0.285 mm of air. The square metal sheets have sides of length 18.6 cm. Calculate the energy stored by the capacitor when a pd of 18.0 V is applied to it. [4]

Good answer (4/4): $C = \dfrac{\varepsilon_0 A}{d}$ and $Q = CV$ so $Q = \dfrac{\varepsilon_0 A}{d}V$

But $U = \dfrac{1}{2}QV$ so $U = \dfrac{1}{2}\dfrac{\varepsilon_0 A}{d}V^2 = \dfrac{1}{2}\dfrac{8.85 \times 10^{-12}\,\text{F m}^{-1} \times (0.186\,\text{m})^2}{0.285 \times 10^{-3}\,\text{m}} \times (18\,\text{V})^2 = 1.74 \times 10^{-7}\,\text{J}$

Bad answer (1/4): $C = \dfrac{\varepsilon_0 A}{d} = \dfrac{8.85 \times 10^{-12} \times 18.6^2}{0.285} = 10.7\,\text{nF}$

$E = \dfrac{1}{2}QV = \dfrac{1}{2}10.7 \times 10^{-7} \times 18 = 97\,\text{nJ}$

The bad answer contains many mistakes. First, the conversions from cm and mm are not carried out. However, the main mistake is a particularly heinous crime. The candidate believes that capacitance, C, from the first equation can be input as the charge, Q, in the second equation. This is quite a common mistake and may have something to do with the capacitance and charge usually having the same SI prefixes (nF, nC or similar).

2. The displacement, x (in metre) of a 56.0 g mass oscillating with SHM is given by the equation:

$$x = 0.250\cos(14.7t)$$

where t is the time in s. Calculate the maximum resultant force acting on the mass. [4]

Good answer (4/4): From looking at the equation, $A = 0.250$ m and $\omega = 14.7\,\text{s}^{-1}$
The maximum acceleration is given by $a_{max} = \omega^2 A$ and we know that $F = ma$.

Combining these, we get $= F_{max} = ma_{max} = m\omega^2 A = 0.056 \times 14.7^2 \times 0.250 = 3.03\,\text{N}$

Bad answer (1/4): I can see that $A = 0.250$ m and $\omega = 14.7$ but the only equation in the booklet says $a = -\omega^2 x$. What time do I use to get ? Help!

The candidate correctly identifies the amplitude and angular velocity but doesn't realise that you must use the amplitude to get the maximum acceleration.

Assessment Objective 3 (AO3)

AO3 questions are ones in which you need to:

analyse, interpret and evaluate scientific information, ideas and evidence, including in relation to issues, to:
1. **make judgements and reach conclusions;**
2. **develop and refine practical design and procedures.**

These questions account for 28% of the marks in the A2 unit papers, 25% in the full A level.

The verbs analyse, interpret and evaluate are all appropriate and this is, indeed, what you will have to do. Most of these AO3 marks will concentrate on the first point – *judgements* and *conclusions*. The context will often be similar to one of the specified practicals with realistic data. Your analysis may well include analysing graphs to make numerical conclusions. You might have to evaluate the quality of the data and your conclusions. In some questions you are given a statement and have to determine whether or not (or to what extent) it is true. There are usually several ways of getting a sensible answer: you must choose one and structure your answer carefully. Other questions relate to the second part of the AO3 statement – develop and refine practical design and procedures. Usually, these questions are based on imperfections in the data and how you could improve the procedure or the apparatus to obtain better data. To answer these questions, you will need to read them carefully because there will be a clue (perhaps right at the start) as to what went wrong.

Another type of question is based on the part of the statement '*including in relation to issues*'. The '**issues**' include, risks and benefits; ethical issues; how new knowledge is validated; how science informs decision making. Try and make sensible comments; the mark scheme will allow for many approaches and the marks will be quite attainable – approach the questions like a politician: have a view. Every theory paper has one issues question.

Examples of AO3 questions

1. According to theory, the temperature of the aluminium block should increase linearly with respect to time. Discuss the quality of the data obtained by Gwydion for the temperature of the aluminium block against time. [4]

Good answer (4/4): At the start of the experiment, there is a slight delay before the temperature of the block starts to increase properly. This is to be expected because the heater is a few cm away from the thermometer. The temperature then rises linearly with time with the straight line of best fit passing through all the error bars until a temperature of 60 °C is reached. After this point, the gradient of the line gradually decreases. This happens because the rate of loss of heat increases as the temperature of the aluminium block increases. Gwydion should have insulated the aluminium block to obtain better results.

Bad answer (1/4): Gwydion's data are of good quality and are accurate because a smooth curve can be drawn through all the data points.

The bad answer does not address the fact that Gwydion's results do not all lie on a straight line.

2. Some people believe that all space exploration is a waste of time and money. Discuss briefly whether, or not, this point of view can be justified. [3]

Good answer (3/3): Some of the most beautiful, artistic images ever produced were actually produced by the Hubble Space Telescope. This type of art, in itself, is worth millions. Sometimes, research must be carried out for the sake of scientific discovery regardless of whether or not a financial gain can be made. Also, the way we're currently killing-off Mother Earth, we might well need somewhere new to live in the future. On the other hand, it is true that billions of dollars have been spent on space-projects that will never save lives or improve the lives of the impoverished or oppressed. However, overall, I would say that the vast majority of people would disagree that space-exploration is a waste of time and money.

Bad answer (1/3): No profits will ever be made out of space exploration so this is a perfectly valid point of view.

The bad answer only makes one point and does not consider the alternative point of view.

Preparing for the examinations

Examination mark schemes

When examiners write questions for A level exams, they also provide mark schemes containing details of how they are to be marked. For an example of a question and its mark scheme, see page 66. You'll notice that each part of the question is covered, with details of the sort of answer required for each mark. The mark scheme also contains information about the assessment objectives and any marks which count towards the mathematical and practical skills on the paper – in this question there are no practical skills. The question is quite heavy on AO3, however.

Let's look at this mark scheme in detail:

Part (a) is a one-mark question which requires both and identification and an explanation. Notice the expression 'or equiv' (*equivalent*) which means that the examiner will look for other ways of explaining which consist of correct physics. All the markers are present or retired physics teachers, so they know how to interpret this.

Part (b) is a typical AO3 question in which you have to draw a conclusion based on data. Notice that the marks are given for the reasons and are only given if the basic conclusion is correct, that it is an example of a strong interaction.

Part (c) is a piece of bookwork, which you are expected to know. Hence it is AO1.

Parts (e)(i) – (iii) are AO2 and include application with calculations. They are really just one question, but they are divided this way to help you pick your way through the different ideas.

In part (e)(iv) notice the letters *ecf*. These stand for *error carried forward*. You have previously calculated the total energy and need to halve it to get the mark here. If your previous calculation gave the wrong answer then you can still pick up mark for using this value. This rule is generally applied even when the mark scheme doesn't say so explicitly.

The marking

Now have a look at some of Rhodri's and Ffion's marked answers to this question. Notice that the examiner has put in ticks and crosses, where the mark has been given or withheld. You'll see also some annotations by the examiner. If you get a mark by ecf, the examiner will write this – see Rhodri's answer to (e)(iv).

In Ffion's (b)(ii) answer, the examiner has written 'not enough for ecf'. This shows that it needed a more specific reason.

Another common annotation is *bod* – see Rhodri's answer to part (c). This stands for *benefit of the doubt*. Rhodri's statement about needing high energy is not explicitly tied in with repulsion but the examiner thought there was enough of a 'hint'.

Notice also that Rhodri used a valid method of calculation which was different from that in the mark scheme in (e)(i). The experienced physics examiner noticed that it was sound physics and awarded the marks.

Unit 3

Containing six topics, Section A of the exam will have around 11 marks per topic, and you might expect the Unit 3 examination to contain six questions – one on each topic. While each year's paper will differ significantly from this basic structure, the examiners will try to distribute their marks equitably between the topics and six questions will be a reasonable rule of thumb. However, there are four things (other than randomness of distribution) that arise to mess up this beautifully symmetric system.

1. Practical content: You can expect 20% of the examination to be based on experimental analysis. This usually means that one (or possibly two) of the questions will be based on one of the six specified practicals for this unit. This could be a description of the method, error analysis, graphs and conclusions – often the longest question on the paper.

2. Quality of extended response (QER): This is a 6-mark question with a lot of lines for writing and maybe some space for diagrams too. These tend to be AO1 marks and so rely on you learning the basic physics required to answer the question. This, however, is only part of the problem. Not only must you put the required information down on paper but you must also do so in a logical, well-presented format, employing good language skills. The penalty for poor spelling, punctuation and grammar is only generally 1 mark at most but the penalty for not knowing the relevant physics is 6 marks! A common type of question for this 6-mark QER is a description of one of the specified practicals.

3. Synoptic content: Although the Unit 3 examination is usually a few days before the Unit 4 examination, you still need to ensure that you have revised Unit 4 thoroughly because of this synoptic content. You should also be familiar with the AS Units 1 and 2. Any of the topics of Unit 1, 2 or 4, can be combined with a Unit 3 topic to make a more difficult question, e.g. nuclear energy or radioactivity could include ideas from the particle physics in Unit 1.

4. Issues: There is always going to be a question about issues, and this will be 2 or 3 AO3 marks. You cannot revise for these questions but do practise the previous questions that have arisen. Just be confident and try to put down some sensible points leading to a sensible conclusion.

Unit 4

Section A of this paper contains five topics leading to a mean of 16 marks per topic and the rule of thumb this time will be five questions. Note that, sometimes, two topics might be combined into one longer question and, likewise one topic might be split into two smaller questions. Everything about practical content, QER, synopticity and issues applies equally to Unit 4 but there is one thing that can be added about the practical content.

Practical content in Unit 4: Examiners will make every effort to ensure that the practical skills examined in Unit 4 are different from those in Unit 3, e.g. log–log graphs will not usually be asked for in both units. The same goes for the other practical skills, such as measuring gradients, describing lines of best fit. So, after Unit 3 you'll know what to look for in Unit 4, but bear in mind that these skills are also tested in the Practical Examination.

Section B has one long question, worth 20 marks, on each of the optional topics. You should answer the question on only one topic.

Key command words and phrases in examination questions

These are the words or phrases which let you know what sort of answer is expected – there are quite a few to look out for:

State: Just provide a statement without an explanation.

Example: State the meaning of A and Z in the $^A_Z X$ symbol for a nuclide.

Answer: Z is the atomic number (or proton number, or number of protons in the nucleus); A is the atomic number (or nucleon number or total number of protons and neutrons in the nucleus).

Define: You need to provide a statement which is close to (or equivalent to) that which appears in the WJEC Terms and Definitions booklet.

Example: Define binding energy of a nucleus.

Answer: It's the energy that has to be supplied in order to dissociate the nucleus into its constituent nucleons.

Explain what is meant by (or explain the meaning of …): This is slightly more complicated than define and can mean a couple of things:

1. Sometimes it just means the same as 'define'
   Example: Explain what is meant by the binding energy of a nucleus.

   Answer: [Exactly the same as above.]

2. Sometimes it's a definition with a number included
   Example: Explain what is meant by the statement, 'The activity of a β^- radioactive source is 1.6 MBq'.

   Answer: The source gives out 1.6×10^6 β^- particles (or electrons) per second.

Explain the difference (between two things): This is two definitions in disguise because if you define both things you have automatically explained the difference between them.

Example: Explain the difference between the magnetic flux and flux linkage through a coil.

Answer: The magnetic flux of a field, B, at an angle θ to a coil of area A is $\Phi = AB\cos\theta$. The flux linkage is the flux multiplied by the number of turns of the coil.

Describe: Provide a brief description but no explanation is required.

Example: Describe how the binding energy per nucleon depends upon the atomic number, A, of nuclei.

Answer: As the nucleon number increases (from 1) the binding energy per nucleon increases (with a spike at ^{4}He) up to a maximum at around ^{56}Fe and then gradually decreases.

Explain … (some statement): Sometimes this requires a reason by reason logical argument. Example: Explain briefly why the fission of ^{235}U releases energy.

Answer: The binding energy per nucleon of ^{235}U is greater than that of the fission products.

Suggest … (or suggest a reason …): Although not a common command word, this can produce some questions that are difficult to answer. These will often be AO3 marks, appearing at the end of a question requiring evaluation skills.

Example: The measured count rate at 10 cm from the radioactive source is much less than that expected from its calculated activity. Suggest a reason.

Possible answers: The radioactive source might be emitting α particles which have a range of less than 10 cm (in air) / the emissions from the middle of the source are absorbed before they are able to emerge.

Calculate or **determine**: The aim is to obtain the correct answer (along with the correct unit, if required by the mark scheme). With this command word, the correct answer will obtain full marks without the workings. However, you are advised strongly to show your working as marks are available for this even if the answer is wrong.

Example: Calculate the magnetic field 1.2 cm from a long straight wire due to a 6.0 A current in the wire.

Answer: $B = \dfrac{\mu_0 I}{2\pi r} = \dfrac{4\pi \times 10^{-7}\ \text{H m}^{-1} \times 6.0\ \text{A}}{2\pi \times 1.2 \times 10^{-2}\ \text{m}} = 1.0 \times 10^{-4}\ \text{T}$

[Note that you do not have to put units in the calculation – but you do in your answer!]

Compare: Not a common command word but you ought to do what it says on the tin – compare the things it says to compare in the question.

Example: The decay constant of nuclide 1 is ten times that of nuclide 2. Compare (a) the half-lives and (b) the activities of samples of the two nuclides with the same number of atoms.

Answer: (a) Half-life of 2 = 10 × half-life of 1. (b) Activity of 1 = 10 × activity of 2.

Evaluate: You will be required to make a judgement, e.g. whether a statement is correct or wrong, or to decide whether data are good or a final value is accurate.

Justify: This is sometimes used in a very similar manner to the word 'determine' when AO3 marks are being examined, e.g. justify whether or not Blodeuwedd was correct in stating that the 2.00 V reading was anomalous.

Discuss: This can often be a command word in the 'issues' question. In general, you will not go far wrong if you make a couple of points in favour of the discussion issue, a couple against it and then draw some sort of a sensible conclusion.

Common exam mistakes

1. **Not converting the given numbers correctly:** Planetary distances are usually in km while wire radii are in mm. Resistors can be in Ω, $k\Omega$, $M\Omega$ and these have to be converted to the correct powers of ten. There are other common conversions such as changing diameter to radius when using area or volume formulae. All these can give rise to simple mistakes which do not show a poor understanding of physics. Such mistakes are not penalised more than one mark most of the time. Nonetheless, these are probably the most common mistakes committed by physics students.

2. **Not reading the question carefully enough**: This usually results in not answering the question that was asked – either by answering a different question altogether or by missing part of the question. The most common parts of questions that are omitted are those that do not have dotted lines for you to answer on, e.g. adding to diagrams. Pay particular attention to these short parts of questions. Other common missed questions are ones that have an **and** condition in the question itself, e.g. calculate the magnitude **and** the direction. One or other part of the question will have been forgotten in the answer.

3. **Not understanding equations properly**. This often involves substituting wrong values into equations – an unforgivable sin! In kinematics equations, for instance, u and v are often mixed up. You shouldn't really have to use the data booklet; you should know the equations intimately and only check it from time to time to ensure that you recollect them correctly. How do you ensure that you don't misunderstand an equation? Practise, practise, practise!

4. **Not knowing the basic terms and definitions** (a surprisingly common cause of loss of marks). There is a WJEC booklet full of these – you should know everything within its covers.

5. **Forgetting to square a value in the equation**. This happens most often with the kinetic energy equation – the equation $E = \frac{1}{2}mv^2$ is written correctly but then the candidate forgets to square the velocity on the calculator. Or the converse: forgetting to square root the answer when using the same equation to calculate the velocity!

6. **Not planning the structure before answering the QER** (and extended explanations). Too many QER responses are rambling and unstructured. This is easily remedied by spending a moment to plan and structure your answer. Using short sentences tends to help, too.

7. **Not matching the correct corresponding values in a calculation**. By far the most common mistake here is with electrical circuits: current, pd and resistance, e.g. a pd and a current will be combined to obtain a resistance ($R = V/I$) but the current and pd do not match – the pd is for one resistor and the current for another.

Unit 3: Oscillations and Nuclei

Section 1: Circular motion

Topic summary

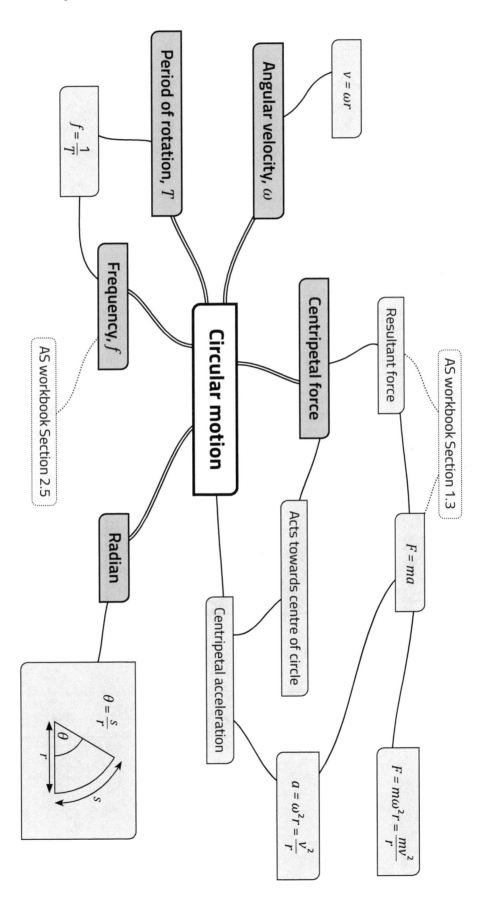

Q1 An angle, θ, has magnitude 1.2 radians. Explain the meaning of this statement, with the aid of a diagram, and explain how an angle in radians can be expressed in degrees (°). [2]

....

....

....

Q2 An object executes circular motion. Define the terms *period of rotation* and *frequency* and explain the relationship between them. [3]

....

....

....

....

Q3 (a) Define *angular velocity* for an object in circular motion. [1]

....

....

(b) The maximum rate of rotation of a washing machine is advertised as 1400 revolutions per minute. Calculate the maximum angular velocity of the washing machine. [2]

....

....

....

Q4 An object of mass 65 kg travels in a circle of radius 4.5 m and at a constant speed, v, of 23.2 m s^{-1} on ice. A light rope connects the object to the centre of the circle.

Calculate the tension in the light rope. [2]

....

....

....

The rope is pulled towards the centre to make it shorter. Explain why this makes the object move more quickly. [2]

....

....

....

....

Q5 A car travels at a constant speed along a flat road and round a curve whose radius of curvature is 24.0 m.

(a) State the direction of the car's acceleration and the resultant force on the car. State what provides this force. [3]

....

....

....

(b) When the magnitude of the centripetal force exceeds the weight of the car, the car will skid and lose control. Calculate the maximum speed that the car can travel around the curve safely. [3]

....

....

....

....

(c) Joe claims that when the radius of curvature of the bend increases, the car can travel faster around the curve but that the maximum angular velocity has to decrease. Determine whether, or not, Joe's claims are accurate. [4]

....

....

....

....

....

....

Q6 (a) Use the data relating to the planet Saturn to answer the following questions:

Planet	Orbit radius (km)	Mass (kg)	Orbit period (year)	Planet radius (km)	Day length (hour)
Saturn	1.43×10^9	5.68×10^{26}	29.5	60 000	10.7

(i) Calculate Saturn's angular velocity about the Sun. [2]

(ii) Calculate Saturn's angular velocity about its North–South axis. [2]

(b) (i) Calculate Saturn's orbital speed (about the Sun). [2]

(ii) Calculate the rotational speed of a point on Saturn's equator. [2]

(c) (i) Calculate the centripetal acceleration and force acting on the planet Saturn and state what provides this force. [4]

(ii) The acceleration due to gravity, g, on the surface of Saturn is 10.4 m s^{-2}. An experiment to measure g at the equator would find a smaller value due to the planet's spin. Calculate the percentage drop in the measured value of g at Saturn's equator. [4]

Q7   The bob of a simple pendulum is made to travel in a horizontal circle, at a constant speed, as shown:

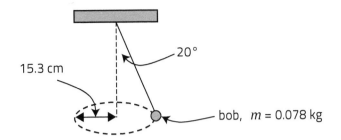

15.3 cm

20°

bob, m = 0.078 kg

(a)   State what provides the centripetal force.   [1]

....

....

....

(b)   By considering the tension in the string, determine the speed of the bob.   [5]

....

....

....

....

....

....

....

....

Q8   (a) Show that the orbital speed, v, of an object in circular orbit of radius, r, about an object of much larger
mass, M, is given by $v^2 = \dfrac{GM}{r}$.   [2]

....

....

....

(b)   Dust at the visible edge of a galaxy has an orbital speed of 4700 km s^{-1}. When the distance from the
centre of the galaxy is doubled, dust has a measured speed of 3400 km s^{-1}. Evaluate whether, or not,
these data disprove the existence of dark matter.   [3]

....

....

....

....

....

....

Question and mock answer analysis

Q&A 1

(a) Explain the difference between velocity and angular velocity for an object in circular motion at constant speed and state the relationship between the magnitudes of the velocity and the angular velocity. [3]

(b) The planet Wolf 1061c is in circular motion of radius 12.6×10^6 km about the star Wolf 1061 and its orbital period is 17.9 days. Calculate:
  (i) the angular velocity of the planet about the star [2]
  (ii) the orbital speed of the planet [2]
  (iii) the centripetal acceleration of the planet [2]
  (iv) the gravitational force exerted by the star Wolf 1061 on the planet Wolf 1061c given that the mass of the planet Wolf 1061c is 2.6×10^{25} kg. [2]

(c) Calculate the mass of the star Wolf 1061. [2]

(d) The star Wolf 1061 is a red dwarf with a low surface temperature and the planet Wolf 1061c has a surface value of g 60% greater than that of Earth. Jemima, a NASA spokesperson, claims that any animals living on Wolf 1061c would be physically strong but that they would not be able to tolerate UV light. Discuss to what extent Jemima might be correct. [3]

What is being asked?

Part (a) is simply a definition question in disguise. If the difference between two terms is required, you'll find that a definition of each term will suffice. Although part (b) looks novel and strange at first sight, it is a very straightforward application of circular motion equations. Part (c) is synoptic because it requires gravitational fields, which is on Unit 4, but you must remember that all A2 papers have a minimum synoptic content. The final part does not seem to be on Unit 3 either and is probably this examiner's attempt at an issue question (probably trying to hit 'Consider applications and implications of science and evaluate their associated benefits and risks.').

Mark scheme

Question part		Description	AOs			Total	Skills	
			1	2	3		M	P
(a)		Velocity, v, is the rate of change of displacement [1]	3			3		
		Angular velocity, ω, is the angle, $\Delta\theta$ (in radian), swept out divided by the time, Δt [1]						
		Relationship $v = r\omega$, where r is radius (of circular motion) [1]						
(b)	(i)	Use of angle in radian divided by time [1]	1	1		2	1	
		$\omega = 4.06 \times 10^{-6}$ rad s^{-1} [1]						
	(ii)	Use of $v = r\omega$ **or** $v = \dfrac{2\pi r}{t}$ [1]	1	1		2	1	
		Correct answer (ecf on ω but needs comment if speed is greater than 3×10^8 m s^{-1}) = 51 200 m s^{-1} [1]						
	(iii)	Use of $a = \omega^2 r$ **or** $a = \dfrac{v^2}{r}$ [1]	1	1		2	1	
		Correct answer (ecf on ω, v and km) = 0.208 m s^{-2} [1]						
	(iv)	Realising gravitational force is the centripetal force (can be implied by answer) [1]		1		2	2	
		Correct answer = ma = 5.41×10^{24} N (ecf on a) [1]		1				
(c)		Rearrangement of $F = \dfrac{GMm}{r^2}$ i.e., $M = \dfrac{Fr^2}{Gm}$ [1]		2		2	2	
		Correct answer = 4.95×10^{29} kg [1]						

| (d) | | Physical strength linked to strong gravity [1] Lower temperature linked to less UV [1] Acceptable conclusion linked to both of the above points or to a new point, e.g. Jemima is wrong because we cannot assume life exists on this planet **or** Jemima's conclusion is correct linked to sensible strength and UV comments [1] | | | 3 | 3 | | |
| **Total** | | | 6 | 7 | 3 | 16 | 7 | |

Rhodri's answers

(a) Velocity is distance over time and angular velocity is angle over time ✓ bod

The equation relating them is $v = \omega r$

MARKER NOTE
Rhodri's definition of velocity is poor – he really should use the word displacement and not distance. His definition of angular velocity is also poor but is far closer to that required in the MS and has been awarded the mark with bod. He has not explained the meaning of r in his equation and cannot receive the last mark.

1 mark

(b)(i) $\dfrac{360}{(17.9 \times 24 \times 3600)}$ X

$= 2.33 \times 10^{-4}\,°s^{-1}$ X

MARKER NOTE
Rhodri has forgotten to use angles in the unit of radian and cannot obtain the 1st mark. There is no ecf within a question part and so he loses the 2nd mark also.

0 marks

(ii) $v = \omega r = 2.33 \times 10^{-4} \times 12.6 \times 10^{6}$ ✓

$= 2935.8$ m/s X no ecf

MARKER NOTE
Rhodri has used the correct equation and gains the 1st mark. He cannot gain the 2nd mark because he has forgotten to convert km to m.

1 mark

(iii) $a = \omega^2 r = (2.33 \times 10^{-4})^2 \times 12.6 \times 10^{6}$ ✓

$= 0.67$ m s^{-2} ✓ ecf

MARKER NOTE
Rhodri has already been penalised for not converting km to m and should not be penalised again. He has also been penalised for his incorrect angular velocity. This means that this answer, with ecf, gains full marks.

2 marks

(iv) $F = ma = 2.6 \times 10^{25} \times 0.67$ ✓

$= 1.7 \times 10^{25}$ N ✓ ecf

MARKER NOTE
Rhodri has used the correct equation and his answer is correct with ecf and so he obtains full marks. Notice that he didn't have to state that the gravitational force was the centripetal force because this is implied in his final answer.

2 marks

(c) I don't understand why this question is on this paper. I wonder what OFQUAL will say about this Mr Examiner?

$M = \dfrac{Gm}{Fr^2}$ X $= 6.4 \times 10^{-25}$ kg X

MARKER NOTE
Rhodri doesn't seem to know that all papers have a synoptic content and so an examiner can ask about Unit 4. Hence this question is fully in line with Ofqual requirements. Rhodri cannot gain any marks because he has rearranged the equation incorrectly and the 1st mark is for correct rearrangement.

0 marks

(d) This is definitely nothing to do with Unit 3. Jemima is probably right because you'd have to be extra strong to cope with the bigger gravity ✓ and they don't sell suntan lotion on Wolf planets.

MARKER NOTE
Once again, Rhodri provides a little light relief for the examiner. He gains the 1st mark for linking extra strength to the stronger gravity. He does not gain the last mark because his UV comment, although funny, is not sensible.

1 mark

Total   **7 marks /16**

Ffion's answers

(a) Velocity is the rate of change of displacement ✓
whereas angular velocity is the rate of change of angle (in rad) ✓
Also, velocity is a vector whereas angular velocity isn't.
The angular velocity is the velocity divided by the radius. ✓

MARKER NOTE
Ffion's definition of velocity is excellent and her definition of angular velocity is, arguably, better than that of the MS. Her fact about angular velocity not being a vector is actually wrong in 3D but this is well beyond the scope of A level and would not be penalised here. Her final relationship is clear with all terms named.

3 marks

(b) (i) $\omega = \dfrac{\theta}{t} = \dfrac{2\pi}{17.9}$ ✓ $= 0.351$ ✗

MARKER NOTE
Although Ffion's answer is a long way from being correct, she is fortunate to obtain the 1st mark because she has used the correct unit for the angle. Her answer is incorrect because she has not converted day to s and does not merit the 2nd mark.

1 mark

(ii) $v = \omega r = 0.351 \times 12.6 \times 10^{6}$ ✓
$= 4.4 \times 10^{6}$ km/s
(oops seems too large) ✗ [not enough for ecf]

MARKER NOTE
Ffion's answer is faster than the speed of light but is correct with ecf. To gain the final mark by ecf she needed to make it clear that it was greater than c.

1 mark

(iii) $a = \omega^2 r = 0.351 \times 12.6 \times 10^{9}$ ✓ bod
$= 4.4 \times 10^{9}$ m s^{-2} ✗ no ecf

MARKER NOTE
Ffion is fortunate to obtain a mark here. Although she has written the correct equation, she has forgotten to square the angular velocity. The examiner has awarded the 1st mark with bod because it seems like she is attempting to use the correct equation.

1 mark

(iv) The gravitational force provides the centripetal force. Hence, ✓
$F = ma = 2.6 \times 10^{25} \times 4.4 \times 10^{9} = 1.1 \times 10^{35}$ N ✓ ecf

MARKER NOTE
Ffion's answer is a long way from the correct answer but each incorrect value that she has used has been punished before. She gains full marks with ecf. Notice how hard the examiners have to work to ensure that ecf is applied correctly!

2 marks

(c) $M = \dfrac{Fr^2}{Gm}$ ✓
$= 1 \times 10^{34}$ kg ✗ no ecf

MARKER NOTE
Ffion obtains the 1st mark for a correct rearrangement but she cannot gain the last mark by ecf because she has forgotten to convert km into m. No ecf is available to Ffion here because this is the first time that she has made this mistake. Again, notice how difficult it is to be an examiner!

1 mark

(d) Extra strength would be useful if g was 16 m s^{-2} because you would have to do more work against gravity when lifting stuff ✓. This star's blackbody spectrum definitely would give less UV because of the lower temperature ✓. However, Jemima's comments might be nonsense because the atmosphere could be carbon dioxide and sulfur dioxide like Venus's.

MARKER NOTE
Ffion's remarks about gravity and UV are to the point and well expressed. Her final comment seems to be about the impossibility of life itself and does not address the question.

2 marks

Total   **11 marks /16**

Section 2: Vibrations

Topic summary

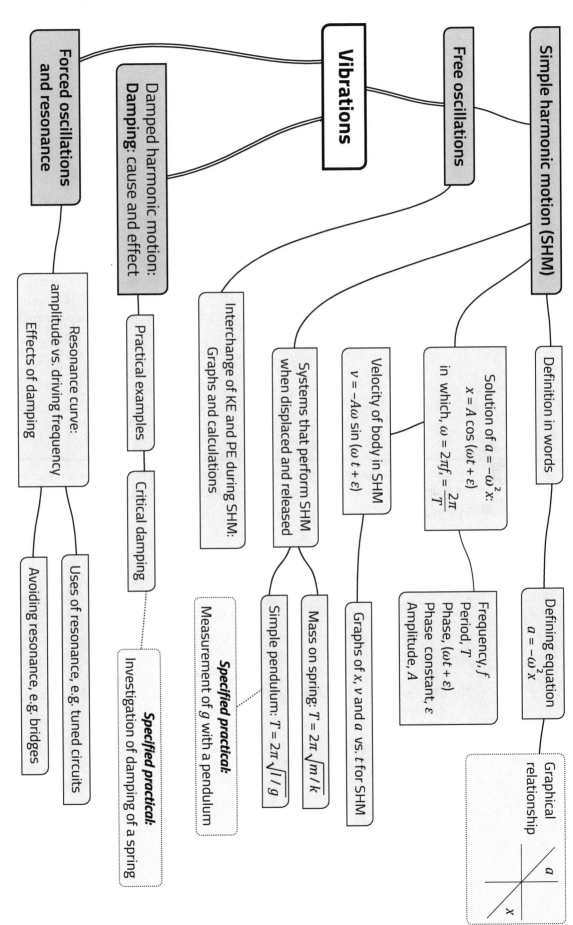

Vibrations

Simple harmonic motion (SHM)

- Definition in words
  - Defining equation $a = -\omega^2 x$
    - Graphical relationship

- Solution of $a = -\omega^2 x$: $x = A \cos(\omega t + \varepsilon)$ in which, $\omega = 2\pi f, = \dfrac{2\pi}{T}$
  - Frequency, f
    Period, T
    Phase, $(\omega t + \varepsilon)$
    Phase constant, ε
    Amplitude, A

- Velocity of body in SHM $v = -A\omega \sin(\omega t + \varepsilon)$
  - Graphs of x, v and a vs. t for SHM

- Systems that perform SHM when displaced and released
  - Mass on spring: $T = 2\pi \sqrt{m/k}$
  - Simple pendulum: $T = 2\pi \sqrt{l/g}$
    - **Specified practical:** Measurement of g with a pendulum

- Interchange of KE and PE during SHM: Graphs and calculations

Free oscillations

Damped harmonic motion: Damping: cause and effect

- Practical examples
  - Critical damping
    - **Specified practical:** Investigation of damping of a spring

Forced oscillations and resonance

- Resonance curve: amplitude vs. driving frequency Effects of damping
  - Uses of resonance, e.g. tuned circuits
  - Avoiding resonance, e.g. bridges

Q1 A displacement–time graph is given on the left for a body performing SHM.

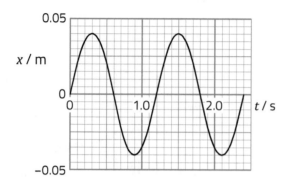 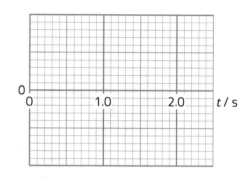

(a) For this oscillation, determine the values of A, ω and ε in the equation

$$x = A \cos (\omega t + \varepsilon).$$

(i) A .... [1]

(ii) ω .... [2]

(iii) ε .... [1]

(b) Using the grid on the right above, sketch a velocity–time graph for the body, providing a vertical scale. Use the space below for working. [4]

Q2 A metal ball attached to a spring whose other end is fixed is given a displacement $x = +0.140$ m from its equilibrium position and released at time $t = 0$. It performs SHM of period 0.800 s. Find:

(a) (i) The ball's displacement at $t = 0.50$ s. [3]

....

....

....

....

....

(ii) The first and second times at which the ball's displacement is +0.070 m. [A rough sketch graph may help.] [3]

....

....

....

....

(b) (i) The ball's velocity at $t = 0.50$ s. [3]

....

....

....

....

(ii) The first and second times at which the ball's velocity is +0.55 m s⁻¹. [A rough sketch graph may help.] [3]

....

....

....

....

....

Q3 The diagram shows a system that performs SHM of frequency 0.40 Hz, in a horizontal plane, when displaced from its equilibrium position and released.

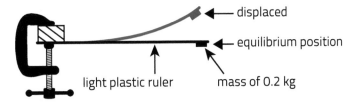

The acceleration, a, of the mass is related to its displacement, x, by the equation:

$$a = \frac{k}{m}x$$

(a) On what part of the system does the value of k depend? [1]

....

(b) Calculate the value of k. [3]

....

....

....

....

....

(c) Sketch a velocity–time (v–t) graph for the first two cycles of the mass's motion, if it is released from rest with displacement $x = +0.050$ m at time $t = 0$. Mark significant values of t and the value of maximum v on the axes. [3]

Space for calculation:

Q4 On the Moon, 100 small oscillations of a simple pendulum take 240 s. On the Earth, 100 small oscillations of the same pendulum take 100 s. Calculate a value for g on the Moon. [3]

Q5 A load of mass 200 g hangs from a spring of stiffness 40 N m⁻¹. The top end of the spring is firmly clamped. The load is pulled down 20 mm below its equilibrium position and then released.

(a) Calculate the time the load takes to reach its highest point. [2]

(b) Fergus says that if the load had been pulled down 30 mm it would have taken a shorter time to reach its highest point, because it would experience a larger resultant upward force. Evaluate Fergus's claim. [3]

Q6 A small object is placed on a spring and the spring extends by a length l. A pendulum is then made of length l (exactly the same as the extension of the spring) and placed to oscillate next to the object on the spring. Davinder notices that both the pendulum and the object on the spring oscillate with exactly the same frequency. Davinder states that this is a complete coincidence. Discuss to what extent Davinder is correct. [5]

Q7 A simple pendulum has a length of 1.00 m. The mass of its bob is 100 g. The pendulum is displaced from the vertical by an angle of 11.5°. The horizontal displacement of the pendulum bob is 0.20 m.

(a) Show that the potential energy of the bob (relative to its lowest point) is approximately 0.02 J. A sketch diagram might help. [3]

....

....

....

....

....

....

(b) The pendulum is released. A graph of (horizontal) displacement, x, against time is given.

On the grids below sketch graphs of:

(i) The pendulum's potential energy, E_p, against time. [3]

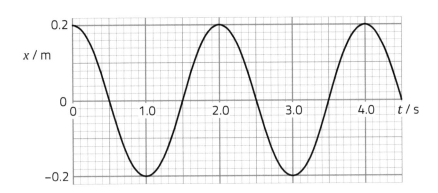

(ii) The pendulum's kinetic energy, E_k, against time. [2]

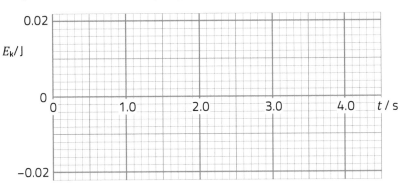

Q8 A velocity–time graph is given for a disc of mass 0.24 kg attached to a spring and oscillating in air.

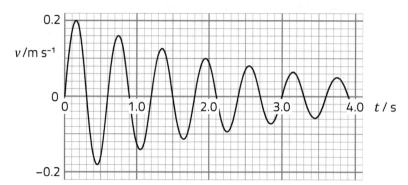

(a) Explain, in terms of forces, why the oscillations are damped. [2]

....

....

....

(b) Sophie believes that the peak values of velocity decrease *exponentially* with time. Evaluate her claim. [3]

....

....

....

....

(c) Determine the percentage of the disc's kinetic energy that is dissipated over the three cycles between the first positive peak and the fourth positive peak. [2]

....

....

....

Q9 (a) State what is meant by *critical damping*. [2]

....

....

....

(b) State one use for critical damping, explaining why lighter damping would not be as suitable. [3]

....

....

....

....

Q10 (a) Define *forced oscillations*. [2]

....

....

....

(b) The diagram shows apparatus that can be used for investigating forced oscillations of a simple pendulum. You are not expected to have seen this apparatus before. [2]

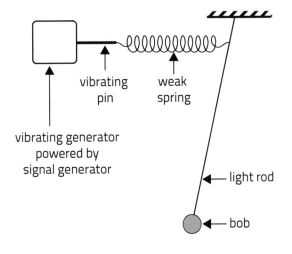
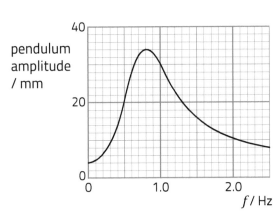

(i) Identify the *driving force* on the pendulum. [1]

....

(ii) Calculate a value for the length of the pendulum, explaining your reasoning. [4]

....

....

....

....

....

....

....

Question and mock answer analysis

Q&A 1

(a) Define *simple harmonic motion (SHM)*. [2]

(b) The diagram shows a system that can perform SHM.

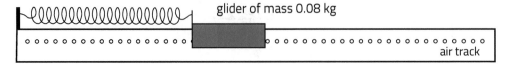

spring (of equal stiffness in compression and extension)

glider of mass 0.08 kg

air track

The glider is displaced along the air track from its equilibrium position and released at time $t = 0$. A displacement–time graph is given on the left tor the glider's motion after release.

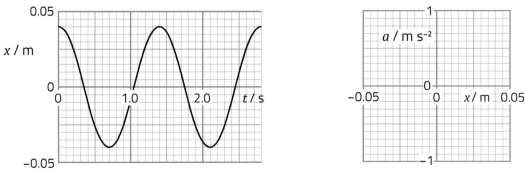

(i) Give the values of the amplitude **and** the period. [2]

(ii) On the grid provided to the right above, draw a graph of acceleration, a, against displacement, x, for the glider. [Space for working is given.] [4]

(iii) Calculate the maximum value of the glider's kinetic energy. [2]

(iv) Ahmed claims that kinetic and potential energy vary at a frequency of 1.43 Hz. Evaluate this claim. [3]

What is being asked

You will have seen the oscillations of a mass hanging from a spring. The apparatus in the question is clearly harder to set up (and a lot more expensive!) but the physics is actually rather easier!

(a) A definition that sets the scene.

(b) (i) No catches, but read the scales carefully!

(ii) Even if you haven't previously met the graph that's asked for, a little thought should tell you the shape. Note that the word 'draw' is used, rather than 'sketch', so your line has to go through the right points.

(iii) This is intended to be a straightforward calculation that shifts the thrust of the question towards the energy aspect of SHM, in preparation for...

(iv) 'Evaluate' is a broad hint that AO3 is being tested. You have to figure out how to approach the evaluation. You could start by mustering what you know about energy in SHM, or you could first try to make sense of the 1.43 Hz.

Mark scheme

Question part		Description	AOs			Total	Skills	
			1	2	3		M	P
(a)		Acceleration proportional to displacement from equilibrium [1] and in opposite direction **or** directed towards equilibrium [1]	2			2		
(b)	(i)	[Amplitude =] 0.040 m [1] [Period =] 1.40 s [1]		2		2		
	(ii)	$\omega = 4.49$ s^{-1} or by implication [1] $\omega^2 = 20.1$ s^{-2} or by implication [1] Line runs between $x = -0.04$ m and $+0.04$ m [1] and between $a = +0.8$ m s^{-2} and -0.8 m s^{-2} with negative gradient [1]		4		4	4	
	(iii)	$v_{max} = 0.040 \times 4.49$ [1] [$= 0.180$ m s^{-2}] or by implication $E_{k\,max} = 1.29$ mJ **unit** [1]		2		2	2	
	(iv)	1.43 Hz shown to be twice the frequency of the displacement variation [1] Any argument that KE varies at this 'double' frequency, e.g. maxima at each pass through equilibrium [1] Any argument that PE varies at this 'double' freq., e.g. maxima at max spring extension and max spring compression [1]			3	3	1	
Total			2	8	3	13	7	0

Rhodri's answers

(a)   Acceleration is proportional to displacement. ✓
      X

MARKER NOTE
Rhodri has left out the statement about the direction. It matters!
1 mark

(b)(i) 0.035 m X   1.40 s ✓

MARKER NOTE
A mistake in reading the amplitude but the period is correct.
1 mark

(ii) $\omega = \dfrac{2\pi}{T} = \dfrac{2\pi}{1.4} = 4.49$ ✓

     max acc $= 4.49 \times 0.035 = 0.16$ X

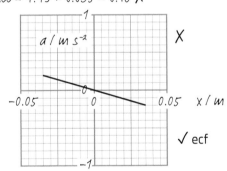

     ✓ ecf

MARKER NOTE
Rhodri gains the 1st mark, for calculating ω. But he hasn't squared it and loses both the 2nd and 4th marks. With ecf on A from (b)(i) he gains the 3rd mark.
2 marks

(iii) $v = -A\omega \sin \omega t$
      Maximum is when $t = 1.05$ s
      $v = -0.35 \times 4.49 \sin(4.49 \times 1.05)$ X
        $= -0.129$
      Max KE $= \dfrac{1}{2} \times 0.08 \times (-0.129)^2$
             $= 6.7 \times 10^{-6}$ J X [no ecf]

MARKER NOTE
Rhodri hasn't seen that the extreme values of the sine function are ±1, so $v_{max} = A\omega$. Instead he's correctly identified the time for maximum v, but has put ωt in radians into a calculator set to degrees. A common mistake, but a costly one.
0 marks

(iv) The frequency is $1/T = 0.714$ Hz ✓
      Ahmed's figure of 1.43 Hz is double this, so Ahmed is wrong.

MARKER NOTE
Rhodri has clearly gained the first mark, but hasn't realised that the variation in energy *should be* at this double frequency.
1 mark

Total   **5 marks /13**

Ffion's answers

(a) The acceleration is proportional to the distance from a fixed point ✓ and is directed towards that point. ✓

> **MARKER NOTE**
> Ffion's statement is correct, though it is safer to use 'displacement' rather than 'distance' (as x in the equation is displacement).
>
> **2 marks**

(b) (i) 0.04 m ✓ 1.4 s ✓

> **MARKER NOTE**
> Both readings correct. It would have been nice to see an extra significant figure given for both values but this was not penalised.
>
> **2 marks**

(ii) $a_{max} = \omega^2 A = \left(\dfrac{2\pi}{1.4}\right)^2 0.04 = 0.81 \ m \ s^{-2}$ ✓✓

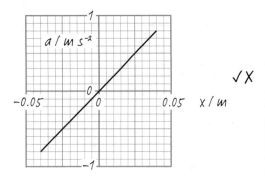

✓ X

> **MARKER NOTE**
> Ffion has done everything correctly except for forgetting the minus sign, so her graph gradient is positive instead of negative. She loses the 4th mark.
>
> **3 marks**

(iii) $v_{max} = A\omega = 0.040 \times \dfrac{2\pi}{1.4} = 0.1795 \ m \ s^{-1}$ ✓

$(KE)_{max} = \dfrac{1}{2} m v_{max}^2 = \dfrac{1}{2} 0.08 \times 0.1795^2$

$= 1.29 \times 10^{-3} \ J$ ✓

> **MARKER NOTE**
> Ffion has used $v_{max} = A\omega$, and has done the calculation correctly.
>
> **2 marks**

(iv) PE is maximum at each extreme of the displacement and KE is a maximum in the middle, which ever way the glider is going. ✓ So the energy is transferred twice as often as the oscillation frequency, and Ahmed is right. ✓

> **MARKER NOTE**
> Ffion understands what is going on, but hasn't actually *shown* that 1.43 Hz is twice the ordinary oscillation frequency, losing the 1st mark.
>
> **2 marks**

| **Total** | **11 marks /13** |

Section 3: Kinetic theory

Topic summary

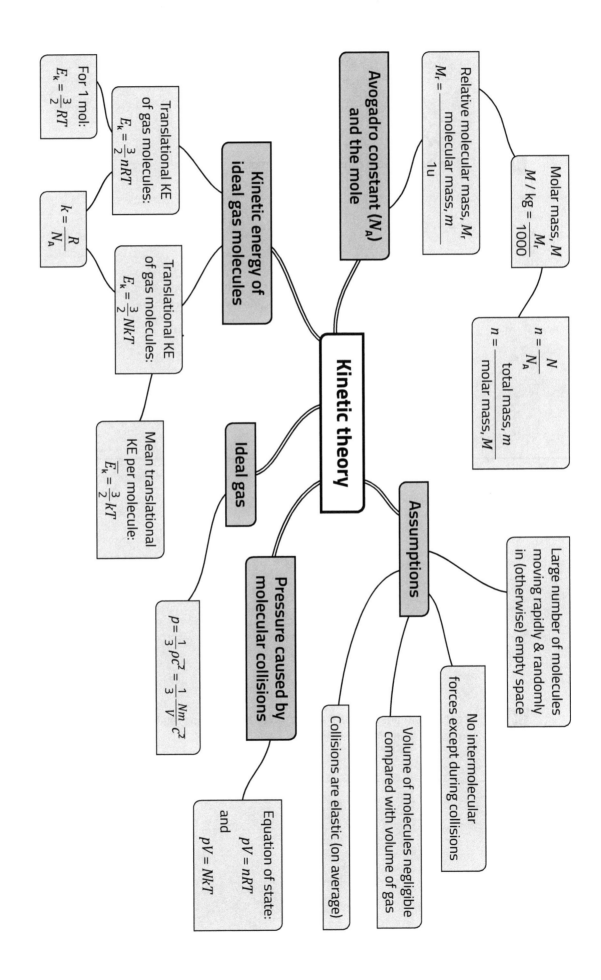

Q1 State the assumptions of the kinetic theory of gases. [4]

Q2 Define the Avogadro constant and the mole. [2]

Q3 Explain, in terms of movement of molecules and Newton's laws, how a gas exerts pressure on its container walls and how this pressure varies with temperature. [6 QER]

Q4 Use the equations $pV = nRT$ and $pV = NkT$ to derive the relationship $k = \dfrac{R}{N_A}$. Make clear what the symbols n, N and N_A represent. [3]

Q5 Use the equations $pV = \frac{1}{3}Nm\overline{c^2}$ and $pV = nRT$ to show that the translational kinetic energy, U, of n mol of monatomic gas is given by $U = \frac{3}{2}nRT$. [3]

....

....

....

....

....

Q6 The explosion inside a car-safety airbag produces 3.0 mol of nitrogen gas (relative molecular mass = 28). The pressure inside the airbag is 140 kPa and the rms speed of the nitrogen molecules is 550 m s⁻¹. Calculate the volume of the airbag. [4]

....

....

....

....

....

....

Q7 A meteorological balloon is released from ground level. The helium in the balloon has an initial volume of 0.89 m³ and a temperature of 298 K. The pressure at ground level is 102 kPa.

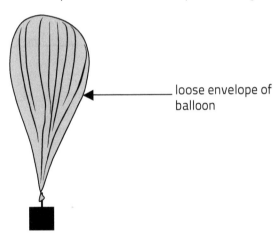

loose envelope of balloon

(a) Calculate the number of molecules of helium in the balloon. [3]

....

....

....

....

....

(b) Calculate the rms speed of the helium molecules (the mass of a helium molecule is 6.64×10^{-27} kg). [2]

....

....

(c) The balloon rises to a height where the pressure is 23 kPa and the temperature is 232 K. Calculate the new volume of the balloon stating any assumption that you make.   [3]

....

....

....

....

Q8   Air is contained in two separate containers connected by a narrow tube fitted with a tap. The air in both containers is in thermal equilibrium with the surroundings, the temperature of which is 293 K.

Volume = 37.0×10^{-3} m^3

Pressure = 1.02×10^5 Pa

Temperature = 293 K

Volume = 22.5×10^{-3} m^3

Pressure = 6.50×10^5 Pa

Temperature = 293 K

The tap is opened and air flows from the right container to the left until the pressures in the two containers are equal and the containers are in thermal equilibrium with the surroundings.

(a) Calculate the final pressure in the containers.   [5]

....

....

....

....

....

....

....

(b) Before thermal equilibrium is reached, Tudor claims that the right container will cool and the left container will become warmer. Discuss whether Tudor is correct.   [3]

....

....

....

....

....

Q9 Air of density 1.35 kg m^{-3} and temperature 293 K, with a pressure of 112 kPa, is trapped in a Cola bottle of volume 1.5 × 10^{-3} m^3.

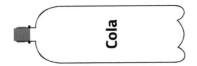

(a) Calculate the rms speed of the air molecules. [3]

(b) Calculate the mass of 1 mol of air. [3]

(c) (i) Air is gradually pumped into the bottle until the bottle explodes when the pressure is 935 kPa and the temperature of the air is 320 K. Calculate the mass of air inside the bottle when this occurs, stating any assumption you make. [3]

(ii) Explain why the temperature inside the bottle increases as air is pumped in. [2]

Question and mock answer analysis

Q&A 1

(a) A sample of an ideal gas, of amount 0.078 mol, at a temperature of 157 K, has a volume of $1.45 \times 10^{-3}\,m^3$.

    (i) Calculate the pressure of the gas.   [2]

    (ii) Explain what is meant by 0.078 mol.   [1]

    (iii) The density of the gas is $4.52\,kg\,m^{-3}$. Calculate the rms speed of the molecules and the molar mass of the gas.   [5]

    (iv) Gwesyn makes two statements about the gas molecules. In each case determine whether the statement is correct:

    Statement 1: 'Halving the mass of each molecule would halve the mean kinetic energy of the molecules (for a given temperature).'

    Statement 2: 'Doubling the temperature would double the rms speed of the molecules.'   [5]

(b)

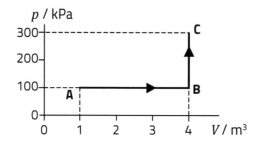

A different sample of a monatomic gas is taken from point A to point C on the $p–V$ diagram. Use calculations to compare:

    (i) the temperatures at A, B and C, and   [2]

    (ii) the ways in which energy is transferred between the gas and its surroundings during the two stages, AB and BC.   [3]

What is being asked?

Part (a)(i) is a simple, one-step calculation; the 1st mark is AO1 but the 2nd mark requiring a calculation is AO2. Part (ii) is related to the definition of the mole (and hence is an AO1 mark). The next two calculations in part (iii) are a little trickier, with the second requiring a two-step calculation. There are a couple of different ways of obtaining the correct answers and these are AO2 marks. Part (iv) is a standard AO3 question in which you have to evaluate statements using physics – with several ways of obtaining the answer. Part (b) overlaps with the Thermal physics area of study, which is quite common as the two have many ideas in common.

Mark scheme

Question part		Description	AOs			Total	Skills	
			1	2	3		M	P
(a)	(i)	Substitution into $pV = nRT$ [1] Answer = 70 200 Pa [1]	1	1		2	1	
	(ii)	N / N_A = 0.078 or equivalent, e.g. 4.7×10^{22} particles [1]	1			1		
	(iii)	Rearrangement of $p = \frac{1}{3}\rho\overline{c^2}$, i.e. $\overline{c^2} = \frac{3p}{\rho}$ (minimum) **or** alternative, e.g. $\frac{3}{2}kT$ and obtaining mass m [1] rms speed = $216\,m\,s^{-1}$ (correct answer only) [1] Total mass = $\rho V = 6.55$ g **or** $m = 1.39 \times 10^{-25}\,kg$ [1] Dividing by the number of moles OR 1u (not both) [1] Answer = $84\,g\,mol^{-1}$ (or $0.084\,kg\,mol^{-1}$) ***unit*** [1]		5		5	5	

	(iv)	**Statement 1**: Kinetic energy only depends on temperature or equation, e.g. $\frac{1}{2}m\overline{c^2} = \frac{3}{2}kT$ [1] Hence Gwesyn is wrong, i.e. correct conclusion linked to correct physics [1] **Statement 2**: KE $\propto T$ or equivalent, e.g. $\frac{1}{2}m\overline{c^2} = \frac{3}{2}kT$ [1] $c_{rms} \propto \sqrt{T}$ or equiv **or** calculation $\longrightarrow c_{rms}$ = 305 m s^{-1} [1] c_{rms} increases by factor $\sqrt{2}$ **or** T must increase by × 4 for it to be true **or** equivalent [1]		5	5	2	
(b)	(i)	Using $pV = nRT$ or $pV \propto T$ [1] $T_C{:}T_B{:}T_A$ = 12:4:1 [1]			2	2	
	(ii)	At least one of: U (A) = 150 kJ, U(B) = 600 kJ and U(C) = 1800 kJ **or** ΔU(AB) = 450 kJ, ΔU(BC) = 1200 kJ [1] W(BC) = 0 **and** W(AB) = 300 kJ [1] Q(AB) = 750 kJ; Q(BC) = 1200 kJ [1]			3	3	
Total			2	6	10	18	8	

Rhodri's answers

(a) (i) $p = \dfrac{nRT}{V} = \dfrac{0.078 \times 8.31 \times 157}{1.45 \times 10^{-3}}$ ✓

= 70 100 Pa ✓ [bod]

MARKER NOTE
Rhodri's working is perfectly correct. He has rounded the pressure (70.18 kPa) incorrectly but this is not always penalised (except in practical questions) – it is correct to 2 sf anyway.
2 marks

(ii) Hah, a doddle, you'll have 0.078 × N_A molecules ✓

MARKER NOTE
Rhodri's answer is equivalent to the mark scheme. Irrelevant comments are ignored unless they are rude or imply a safeguarding issue.
1 mark

(iii) $\overline{c^2} = \dfrac{3p}{\rho}$ ✓ = 46 500 m s^{-1}

$M = \rho V = 0.006554$ ✓

Mm = 0.006554 / (0.078 × 6.02 × 10^{23})

Mm = 1.396 × 10^{-25}

MARKER NOTE
Rhodri has made a common mistake. He has calculated the mean square speed and forgotten to take the square root.
Rhodri's 2nd answer is also incorrect because he has calculated the mass of a molecule rather than a mole. Mixing up the various masses is a common mistake. Note that Rhodri cannot obtain the penultimate mark because he has divided by both 0.078 and N_A.
2 marks

(iv) Since KE = mc^2

It's obvious that KE is proportional to mass and Gwesyn is right ✗

MARKER NOTE
Rhodri has fallen into the trap here and has forgotten that lighter particles will have greater rms speeds at the same temperature. He has simply looked at the kinetic energy formula and come to the wrong conclusion.
0 marks

Using 1/2mc^2 = 3/2 kT ✓

So c^2 is proportional to T

And Gwesyn is correct (any chance of more difficult questions next time please ☺)

MARKER NOTE
Rhodri gains the 1st mark because he realises that $\frac{1}{2}m\overline{c^2} = \frac{3}{2}kT$ is a suitable starting point. In spite of his smugness, he then makes the same mistake as in part (c) – considering the mean square speed and not the rms – an error of <u>physics</u>, so no ecf will be awarded here.
1 mark

(b)(i) $pV = nRT$ so $T = \dfrac{pV}{nR}$ ✓

But we don't know n so we can't work out the temperature – not enough information!

MARKER NOTE
Rhodri gains a mark because $pV = nRT$ is a good starting point. He doesn't realise the significance of the instruction to <u>compare</u> the temperatures, e.g. finding the ratio, so he cannot access the 2nd mark.
1 mark

(ii) A to B: W = area under graph

= 3 × 100 × 10^3

= 300 000 J

B to C: W = 0 (volume constant) ✓

MARKER NOTE
The only mark that Rhodri accesses is the one for the two values of Q at all, probably because he has no idea how to calculate ΔU – which is similar to his difficulty with part (i).
1 mark

Total   **8 marks /18**

Ffion's answers

(a) (i) $p = \frac{nRT}{V} = 70$ kPa ✓✓

MARKER NOTE
Ffion has rearranged the equation correctly and has obtained the correct answer for full marks. Ffion has rounded correctly to 2 sf here which is ideal since the number of moles is only to 2 sf. Note that sfs and rounding are mainly penalised in the practical questions. **2 marks**

(ii) The no. of particles is 0.078 × the no. of particles in 12g of carbon-12 ✓

MARKER NOTE
Ffion's answer is equivalent to Rhodri's but she has also provided an (unnecessary) out-of-date definition of a mole (it is now defined as $6.022\ 140\ 76 \times 10^{23}$ particles but this happened after the Terms & definitions booklet was written). **1 mark**

(iii) $\frac{1}{2}m\overline{c^2} = \frac{3}{2}kT$: First calculate m ✓

$M = \rho V = 4.52 \times 0.00145 = 0.006554$ kg

$m = \dfrac{6.554 \times 10^{-3}}{0.078 \times 6.02 \times 10^{23}} = 1.396 \times 10^{-25}$ kg

$\overline{c^2} = \dfrac{3kT}{m}$, so $c_{rms} = 216$ m s^{-1} ✓

Molar mass $= \dfrac{1.396 \times 10^{-25}}{1.66 \times 10^{-27}} = 87$ ✓✓unit

MARKER NOTE
Ffion's answer is sound but she has made life difficult for herself by taking the alternative route to finding the rms speed as well as an alternative method of finding the molar mass. Also the two calculations are interwoven – a problem for the examiner! Her only omission is the final unit for the molar mass – and the other method, i.e. dividing the total mass by the number of moles makes this a less likely mistake. **4 marks**

(iv) Using $\frac{1}{2}mc^2 = \frac{3}{2}kT$ again ✓
The right-hand side stays the same (same temp) so the left-hand side stays the same and so Gwesyn is wrong ✓

MARKER NOTE
Ffion's answer has made a nice (concise) answer which is fully correct – gaining full marks. **2 marks**

Using $\frac{1}{2}m\overline{c^2} = \frac{3}{2}kT$ ✓
If c ×2 then c² ×4 ✓ So you would need to quadruple the temperature to double the speed and Gwesyn is wrong again. ✓

MARKER NOTE
Ffion has chosen to double the (rms) speed and show that the temperature would be multiplied by 4. An equally valid way would be to double T and show that the speed is multiplied by $\sqrt{2}$. Anyway – a perfect answer and full marks. **3 marks**

(b) (i) $pV \propto T$.
Point A: $pV = 100k \times 1 = 100\ 000$ ✓
Point B: $pV = 400\ 000$
Point C: $pV = 1\ 200\ 000$
∴ $T_C = 12 \times T_A$ and $T_B = 4 \times T_A$ ✓

MARKER NOTE
Ffion uses the relationship $pV \propto T$ and makes no attempt to calculate the actual temperatures but is able to find the factors by which B and C are above the temperature of A – picking up both marks. **2 marks**

(ii) $A \longrightarrow B: W = 100 \times (4 - 1) = 300$ kJ
$U = \frac{3}{2}nRT = \frac{3}{2}pV$
∴ $\Delta U = \frac{3}{2} \times 100 \times (4 - 1) = 450$ kJ ✓
∴ $Q = \Delta U + W = 750$ kJ
$B \longrightarrow C: W = 0$ ✓
$\Delta U = \frac{3}{2} \times (300 - 100) \times 4 = 800$ kJ ✗
∴ $Q = 1200$ kJ

MARKER NOTE
Ffion's answer is almost perfect. She doesn't calculate the internal energy at any of A, B or C but calculates the ΔU(AB) correctly (1st mark). W for the two stages gains her the 2nd mark. She cannot obtain the third mark because she makes a mistake in calculating ΔU(BC) – forgetting to use the factor of $\frac{3}{2}$ **2 marks**

Total **16 marks /18**

Section 4: Thermal physics

Topic summary

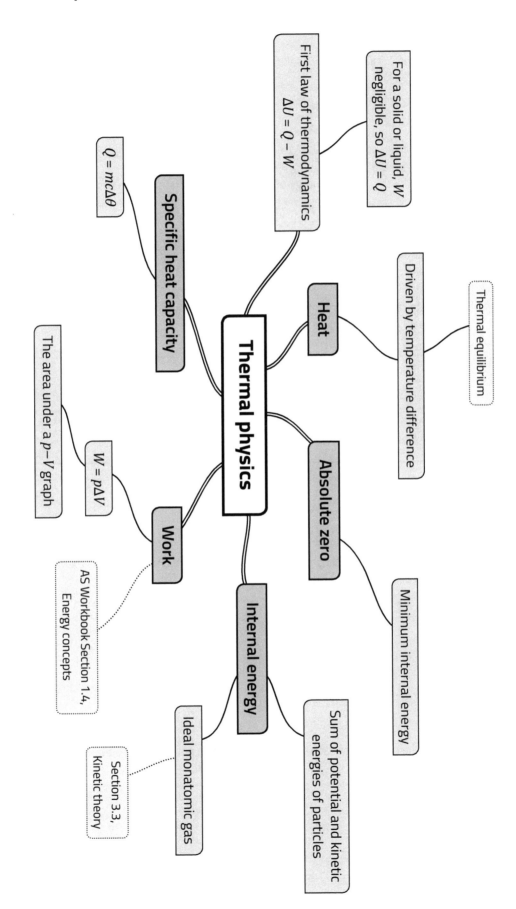

First law of thermodynamics
$\Delta U = Q - W$

For a solid or liquid, W negligible, so $\Delta U = Q$

$Q = mc\Delta\theta$

Specific heat capacity

Thermal physics

Heat

Driven by temperature difference

Thermal equilibrium

Absolute zero

Minimum internal energy

The area under a p–V graph

$W = p\Delta V$

Work

AS Workbook Section 1.4, Energy concepts

Internal energy

Ideal monatomic gas

Section 3.3, Kinetic theory

Sum of potential and kinetic energies of particles

Q1 State what is meant by the *internal energy of a system*. [2]

....

....

....

Q2 Explain the significance of *absolute zero* with regards to internal energy. [2]

....

....

....

Q3 (a) Explain why the internal energy of an ideal gas is different from that of systems in general. [2]

....

....

....

(b) Calculate the internal energy of 30 g of neon gas at a temperature of 26.85 °C (the molar mass of neon is 20 g). [2]

....

....

....

Q4 Explain what is meant by the term *heat*. [2]

....

....

....

Q5 Two systems in thermal contact are in *thermal equilibrium*. State what *thermal equilibrium* means in terms of *heat* and *temperature*. [2]

....

....

....

Q6 (a) The first law of thermodynamics can be written in the form:

$$\Delta U = Q - W$$

Explain the meaning of each term in the equation and how this equation represents conservation of energy. [3]

....

....

....

....

....

(b) Explain why the first law of thermodynamics reduces to $\Delta U = Q$ for a solid or liquid. [2]

....

....

....

Q7 Define the *specific heat capacity* of a substance. [2]

....

....

....

Q8 A group of students uses the following apparatus to investigate how the pressure of a sample of air varies with temperature, at constant volume. Describe how they could use the apparatus to obtain an estimate of absolute zero. [6 QER]

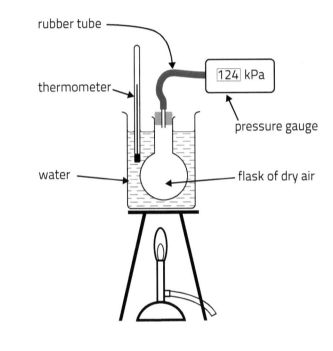

....

....

....

....

....

....

....

....

....

....

....

Q9 A gas is expanded quickly so that no heat is transferred to the gas. Explain why the temperature of the gas decreases. [3]

Q10 (a) The volume of a gas is increased by $2.7 \times 10^{-3}\,m^3$ at a constant pressure of $1.42 \times 10^5\,Pa$. Calculate the work done by the gas. [2]

(b) The same gas is then compressed from this increased volume at the constant pressure of $1.42 \times 10^5\,Pa$ to a final volume which is $1.5 \times 10^{-3}\,m^3$ less than the original volume at the start of part (a). Calculate the work done by the gas. [3]

Q11 Carrots of mass 0.700 kg and specific heat capacity $1880\,J\,kg^{-1}\,K^{-1}$ at a temperature of 20°C are placed in 1.2 kg of boiling water. The specific heat capacity of water is $4210\,J\,kg^{-1}\,K^{-1}$.

Calculate the equilibrium temperature of the carrots and water, stating any assumptions that you make. [5]

Q12 A sample of gas is taken around a closed cycle ABCA as shown in the p–V diagram.

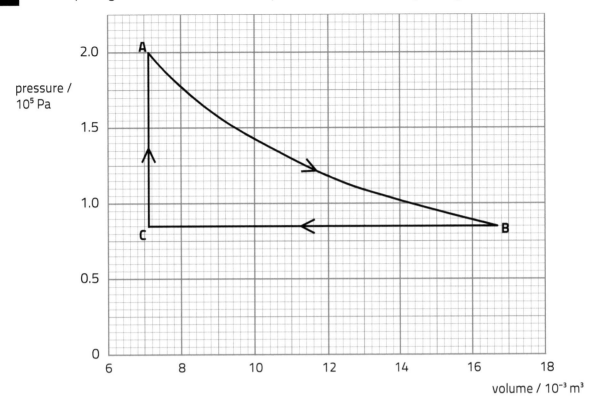

(a) Calculate the work done by the gas:

(i) for process CA; [1]

....

(ii) for process BC; [2]

....

....

....

(iii) for process AB. [3]

....

....

....

....

....

(b) Charlie claims that process AB is an isotherm, i.e. it occurs at a constant temperature. Discuss why Charlie is correct. [4]

....

....

....

....

....

....

(c) Complete the following table with the correct data: [6]

Space for calculations:

	AB	BC	CA	ABCA
ΔU / J	0			
Q / J				
W / J	(a)(iii)	(a)(ii)	(a)(i)	

Q13 Tegfryn carries out an experiment to measure the specific heat capacity of aluminium using the standard apparatus shown below.

He uses no insulation and the temperature of the block starts from room temperature of 20°C. He also measures the following values:

Mass of aluminium block = 1.000 kg, pd = 12.00 V, current = 4.20 A.

He checks the internet and finds that the specific heat capacity of aluminium is 900 J kg^{-1} K^{-1}.

Plot a graph of the expected results on the grid below. [6]

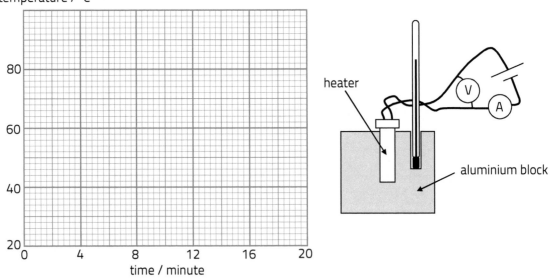

Space for calculations:

Q14 (a) Show that the equation for the work, W, done by an expanding gas

$$W = p \, \Delta V$$

is correct as far as units (or dimensions) are concerned. [2]

....

....

....

(b) A syringe with its outlet blocked contains 110×10^{-6} m³ of argon (a monatomic gas) at a temperature of 20 °C. The syringe is immersed in boiling water at 100 °C. The argon expands to a volume of 140×10^{-6} m³, by moving the piston of the syringe. The pressure is 100 kPa throughout.

(i) Verify that a negligible amount of gas escapes during the expansion. [4]

....

....

....

....

....

(ii) **Giving your reasoning clearly**, show that 7.5 J of heat enters the gas during the expansion. [5]

....

....

....

....

....

....

(iii) Lucia uses the data in this question to reach the following conclusion: 'The heat needed to raise the temperature of 1.0 mol of argon gas by 1.0 K is 21 J.' Examine to what extent her statement is justified. [4]

....

....

....

....

....

....

Question and mock answer analysis

Q&A 1

(a) The first law of thermodynamics can be written in the form:

$$\Delta U = Q - W$$

State the meaning of the three terms: ΔU, Q and W. [3]

(b) An ideal monatomic gas goes through the cycle ABCA shown in the graph:

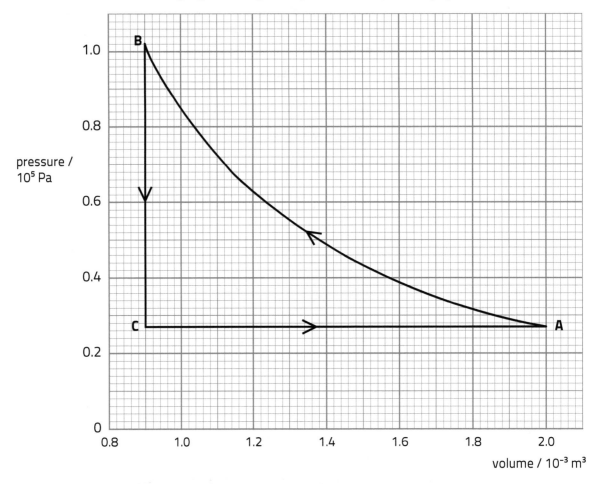

(i) The temperature at A is 293 K. Calculate the number of moles of gas. [2]

(ii) Calculate the temperatures at points B and C. [3]

(iii) Calculate the change in internal energy between A and B. [2]

(iv) Show that the work done by the gas during part AB is approximately −60 J. [3]

(v) Teilo states that, during process AB, no heat is transferred. Determine to what extent Teilo is correct. [2]

(vi) Calculate the values of ΔU, Q and W for the whole cycle ABCA. [4]

What is being asked?

Part (a) is the gentle introduction with three very easy AO1 marks. Part (b) is mainly analysis of a gas cycle. Parts (i) & (ii) are related to the previous topic (kinetic theory) but you can expect a mix 'n' match of these two topics often. Nonetheless, these are not difficult calculations – they involve one equation only ($pV = nRT$). Part (iii) is based on this topic but is also a simple one-step calculation. Part (iv) is a little more complicated and requires an area below the curve to be approximated. Whereas (i)–(iv) are mainly AO2 skills, the wording of (v) is typical of AO3 marks. There are many ways to complete part (v), as is usually the case with these questions. The final part tests AO2 skills and requires a good understanding of both a closed cycle and the first law of thermodynamics.

Mark scheme

Question part		Description	AOs 1	AOs 2	AOs 3	Total	Skills M	Skills P
(a)		ΔU, change in internal energy [1] Q, heat supplied to gas/system [1] W, work done by gas/system [1]	3			3		
(b)	(i)	Use of $pV = nRT$ [1] $n = 0.0222$ mol [1]	1	1		2	1	
	(ii)	Either use of $pV = nRT$ or pV/T = constant [1] $T_B = 498$ K [1] $T_C = 132$ K [1]	1	2		3	2	
	(iii)	Use of $U = \frac{3}{2}nRT$ or $\frac{3}{2}pV$ [1] Answer = 57 J (ecf) [1]	1	1		2	1	
	(iv)	Application of work = area under curve [1] Realisation of work done on gas or −ve sign linked to compression [1] Accurate answer shown [accept 52 J to 64 J] [1]		3		3	3	
	(v)	Application of first law, e.g. $\Delta U = Q - W$ $60 = 0 - (-60)$ OR $Q = \Delta U + W = 60 - 60 = 0$ [1] valid conclusion in light of first law (ecf) [1]			2	2	1	
	(vi)	$\Delta U = 0$ (start & end temps are equal) [1] Area for CA calculated or implied e.g. $\Delta V \times p = 1.1 \times 10^{-3} \times 0.27 \times 10^5$ (29.7 J) [1] hence $W = -30$ J [1] $Q = W$ [1]		4		4	2	
Total			6	11	2	19	10	

Rhodri's answers

(a) ΔU is internal energy X (not enough)

Q is heat going in or out of the gas X

W is work done on or by the gas X

MARKER NOTE
Each of Rhodri's answers falls short of the mark scheme and he gains no marks at all. He is not far from gaining 3 marks but each statement is slightly wrong or ambiguous.
0 marks

(b)(i) $n = \dfrac{pV}{RT} = \dfrac{0.27 \times 2}{8.31 \times 293}$ ✓ $= 0.000222$ mol X

MARKER NOTE
Rhodri gains the 1st mark but not the 2nd because he has failed to notice the correct multipliers in the units on the graph.
1 mark

(ii) $T = \dfrac{pV}{nR} = \dfrac{12 \times 0.9}{0.000222 \times 8.31} = 585$ ✓

$T = \dfrac{pV}{nR} = \dfrac{0.27 \times 0.9}{0.000222 \times 8.31} = 132$ ✓

MARKER NOTE
Rhodri has applied the equation correctly for the 1st mark and has obtained the correct answer for the 3rd mark (notice how the powers of 10 slips out in this section). Rhodri does not gain the 2nd mark because he has read the pressure scale incorrectly (the pressure should be 1.02×10^5 Pa).
2 marks

(iii) $U = \frac{3}{2}nRT = 1.5 \times 0.000222 \times 8.31 \times 585$ ✓

$= 1.08$ J X

MARKER NOTE
Here, Rhodri has not calculated the change in internal energy, only the internal energy at the (wrong) higher temperature. Nonetheless, he has used the equation and gains a generous mark.
1 mark

(iv) Counting squares, I guess there are 13.5 large squares below the curve ✓

Each square is $0.2 \times 0.2 = 0.04$ J

Work done $= 0.04 \times 13.5 = 0.54$ J

MARKER NOTE
Rhodri's final answer is correct after he has multiplied his answer by a factor of 100. His explanation of why he has done this is given bod by the examiner. He has also stated that the work is done on the gas.
3 marks

But pressure is $\times 10^5$, volume $\times 10^{-3}$ so work done is 54 J. Also, this work is being done on the gas not by it. $\checkmark\checkmark$ bod $\times 100$ (ecf)

(v) $\Delta U = 1.08$J and $W = -54$ J

$\Delta U = Q - W$ so $Q = \Delta U + W = -53$ J $\checkmark$

So it seems to me that Teilo is no Einstein and he has got it wrong. $\checkmark$ ecf

MARKER NOTE
Rhodri's previous ΔU is wrong but he cannot be penalised for this again. He has found a good method of checking, i.e. applying the 1st law and his conclusion is correct with ecf.

2 marks

(vi) work done for CA $= 0.27 \times 1.1 = 0.297$ J $\checkmark$ bod

So work done $= -54 + 29.7 = -24.3$J $\checkmark$

I suppose I'm meant to use $\Delta U = Q - W$ but I can't work out $\Delta U = Q - W$

MARKER NOTE
Rhodri's workings for the work done are correct. His numbers are not exactly the same as the MS but within tolerance. He cannot gain any more marks because he does not realise that $\Delta U = 0$ for a closed cycle.

2 marks

Total   **11 marks /19**

Ffion's answers

(a) ΔU is the increase in internal energy $\checkmark$

Q is heat entering the system $\checkmark$

W is work done by the gas $\checkmark$

MARKER NOTE
Ffion's answers are excellent, and she gains full marks. Note that 'change' is always defined as 'final − initial' so that change and increase are equivalent. Ffion is inconsistent in her use of 'system' and 'gas' but this is not penalised.   **3 marks**

(b)(i) $n = \dfrac{pV}{RT} = \dfrac{0.27 \times 10^5 \times 2.00 \times 10^{-3}}{8.31 \times 293}$ $\checkmark$

$= 0.0222$ mol $\checkmark$

MARKER NOTE
Ffion's answer is correct and she gains both marks.   **2 marks**

(ii) $\dfrac{p_A V_A}{T_A} = \dfrac{p_B V_B}{T_B}$, $\checkmark$ so

$T_B = \dfrac{p_B V_B}{p_A V_A} T_A = \dfrac{1.05 \times 0.90 \times 293}{0.27 \times 2.00} = 513$ K

$T_C = \dfrac{V_C}{V_A} T_A = \dfrac{0.90 \times 293}{2.00} = 132$ K

and $T_C = 132$ K $\checkmark\checkmark$

MARKER NOTE
Ffion uses the $\dfrac{pV}{T}$ = constant form of the ideal gas law. She gains full credit for this. Note that she doesn't need to use the multipliers because they cancel when the ratios of the pressures and volumes are used in the calculations.   **3 marks**

(iii) $\Delta U = \dfrac{3}{2} nR (513 - 293) = 61$ J $\checkmark\checkmark$

MARKER NOTE
Ffion's answer gains full marks even though it is wrong at first glance. This is where the examiner has to check Ffion's numbers from a previous answer and find that she has done everything correctly.   **2 marks**

(iv) Approximating AB to a straight line

Area $= 0.5 \times (1.02 + 0.27) \times 10^5 \times 1.1 \times 10^{-3}$

$= 71$ J $\checkmark$

Which is close to 60 J   QED

MARKER NOTE
Ffion's approximation of the area is too large – she has approximated AB to a straight line. She has not explained the negative sign either and only gains the 1st mark for knowing that the area under the curve is the work.   **1 mark**

(v) in $\Delta U = Q - W$, we have

$61 = Q + 71$ So $Q = 10$ J $\checkmark$ bod

And the heat transferred is small in comparison so Teilo is quite accurate. $\checkmark$

MARKER NOTE
The examiner could have penalised Ffion because her previous W has changed sign. Bod was awarded because the examiner believes that Ffion is correcting her previous mistake with the sign.   **2 marks**

(vi) $\Delta U = 0$ because it's a closed cycle $\checkmark$

WD for CA$= 0.27 \times 10^5 \times 1.1 \times 10^{-3} = 30$ J $\checkmark$

So total WD $= 30 - 70 = 40$ J $\checkmark$ ecf

This is also the heat because $Q = W$ if internal energy doesn't change $\checkmark$

MARKER NOTE
Ffion's answer is perfect even though her final answer is a little large. This is just a knock-on effect from part (iv) and she deserves full marks with ecf.   **4 marks**

Total   **17 marks /19**

Section 5: Nuclear decay

Topic summary

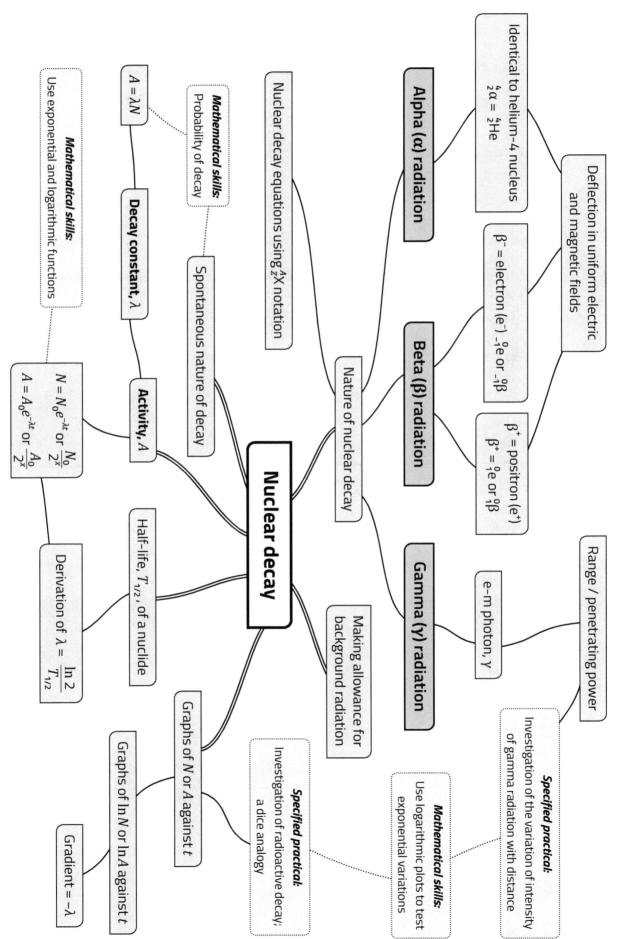

Nuclear decay

Nature of nuclear decay

Alpha (α) radiation
- Identical to helium-4 nucleus $_2^4\alpha = {}_2^4\text{He}$
- Deflection in uniform electric and magnetic fields

Beta (β) radiation
- $\beta^- = $ electron (e$^-$) $_{-1}^0$e or $_{-1}^0\beta$
- $\beta^+ = $ positron (e$^+$) $\beta^+ = {}_1^0$e or $_1^0\beta$
- Deflection in uniform electric and magnetic fields

Gamma (γ) radiation
- e-m photon, γ
- Range / penetrating power

Specified practical:
Investigation of the variation of intensity of gamma radiation with distance

Nuclear decay equations using $_Z^A$X notation

Spontaneous nature of decay

Mathematical skills:
Probability of decay

$A = \lambda N$

Decay constant, λ

Activity, A

$N = N_0 e^{-\lambda t}$ or $\dfrac{N_0}{2^x}$

$A = A_0 e^{-\lambda t}$ or $\dfrac{A_0}{2^x}$

Mathematical skills:
Use exponential and logarithmic functions

Derivation of $\lambda = \dfrac{\ln 2}{T_{1/2}}$

Half-life, $T_{1/2}$, of a nuclide

Making allowance for background radiation

Graphs of N or A against t

Graphs of $\ln N$ or $\ln A$ against t

Gradient $= -\lambda$

Specified practical:
Investigation of radioactive decay: a dice analogy

Mathematical skills:
Use logarithmic plots to test exponential variations

Q1 When asked to describe what β radiation was, Alex wrote, 'It is a stream of electrons.' Charlie said that Alex's description was incomplete.

State what is missing from her description. [2]

....

....

....

Q2 The **activity**, A, of a radioactive sample can be calculated using the equation:

$$A = \lambda N$$

(a) State what is meant by the *activity* and give its unit. [2]

....

....

....

(b) The decay constant of plutonium-239 is 9.11×10^{-13} s^{-1}. A student reads that, in any year, an atom of plutonium-239 has a probability of less than 1 in 30 000 of decaying.

Evaluate whether this is correct. [3]

....

....

....

....

....

Q3 In an experiment to investigate the radiation emitted by a radioactive sample, a student set it up 10 cm from a radiation detector. Counts were taken over 5-minute periods with no absorber, a thin paper absorber and a 3 mm aluminium absorber. The results were as follows:

Absorber	none	paper	aluminium
Count	576	570	568

(a) Explain what conclusions can be drawn from the results about the emission of α, β and γ radiation by the sample. [3]

....

....

....

....

....

(b) Suggest two improvements to the experiment and explain how they will allow more complete conclusions to be drawn. [3]

....

....

....

....

Q4 Uranium-235 ($^{235}_{92}U$), decays by α-emission to an isotope of thorium (Th).

(a) Write the nuclear decay equation for $^{235}_{92}U$. [3]

$$^{235}_{92}U \longrightarrow$$

(b) The thorium isotope subsequently undergoes a series of α and β⁻ decays until a stable isotope of lead is formed. There are three stable isotopes of lead, $^{206}_{82}Pb$, $^{207}_{82}Pb$ and $^{208}_{82}Pb$.

(i) Paul says that the isotope formed must be $^{207}_{82}Pb$. Explain why he is correct. [2]

....

....

(ii) Determine the number of α decays and β⁻ decays that take place in the decay series from $^{238}_{92}U$ to $^{206}_{82}Pb$ and explain your answer. It might help you to complete the following equation (you may ignore neutrinos): [3]

$$^{238}_{92}U \longrightarrow {}^{206}_{82}Pb + \text{.....................} + \text{.....................}$$

....

....

....

....

(iii) The uranium isotope $^{233}_{92}U$ is part of a different radioactive decay series. Explain why this series cannot **end** on an isotope of lead. [2]

....

....

....

Q5 A teacher sets up a cobalt-60 gamma source 20 cm from a radiation detector and rate meter. The reading, corrected for background, is 9.76 counts per second. Exactly one year ago, the reading was 11.50 counts per second.

(a) Calculate what the reading will be:

(i) one year from now. [2]

....

....

(ii) in 10 years' time. [1]

....

(b) The level of background radiation in the teacher's lab is 0.42 counts per second. Determine the time it will take for the cobalt-60 source to decay to the same level as this measured at 20 cm distance. [3]

....

....

....

Q6 High speed particles from space, known as cosmic rays, collide with atoms in the upper atmosphere. Some of the collisions result in the emission of neutrons which are absorbed by nitrogen, $^{14}_{7}N$, nuclei, producing $^{14}_{6}C$ nuclei, which are β^{-} radioactive with a half-life of 5730 years. The balance, between the production of $^{14}_{6}C$ and its decay, results in a ratio of $^{14}_{6}C / ^{12}_{6}C$ atoms of 1.250×10^{-12}.

The tissues of living organisms contain $^{14}_{6}C$ in the same ratio to $^{12}_{6}C$ as in the atmosphere. After an organism dies, the level of $^{14}_{6}C$ in its tissues decreases, due to radioactive decay. This decrease can be used to estimate the age of objects made from biological materials in a process called radio-carbon dating.

(a) Complete the nuclear equation for the production of $^{14}_{6}C$. [2]

$$\underline{}n + {}^{14}_{7}N \longrightarrow {}^{14}_{6}C +$$

(b) Write the nuclear decay equation for $^{14}_{6}C$. [2]

$$^{14}_{6}C \longrightarrow \text{.................} + \text{.................} + \text{.................}$$

(c) (i) Calculate the radioactive decay constant, λ, for carbon-14. [2]

....

....

....

(ii) A wooden artefact from ancient Egypt was found to have a $^{14}_{6}C / ^{12}_{6}C$ ratio of $(0.851 \pm 0.002) \times 10^{-12}$. Estimate the age of the wood together with its absolute uncertainty. [3]

....

....

....

....

....

(iii) The burning of fossil fuels in the last 200 years has reduced the amount of $^{14}_{6}C$ in the atmosphere by 3%, compared to $^{12}_{6}C$. Sioned says that this will make recently manufactured objects seem older than they really are, if assessed using radio-carbon dating. Evaluate whether Sioned is correct. [2]

....

....

....

....

....

Q7 A class of students investigates radioactive decay theoretically, by imagining an experiment involving 800 octahedral dice, each of which has one face painted black.

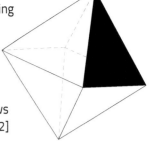

The students consider throwing the dice several times and, each time, removing any that land with the black face up.

(a) Show that the number of dice the students expect to remain after n throws is $800 \times (0.875)^n$. [2]

....

....

....

(b) The students then perform the experiment with the following results:

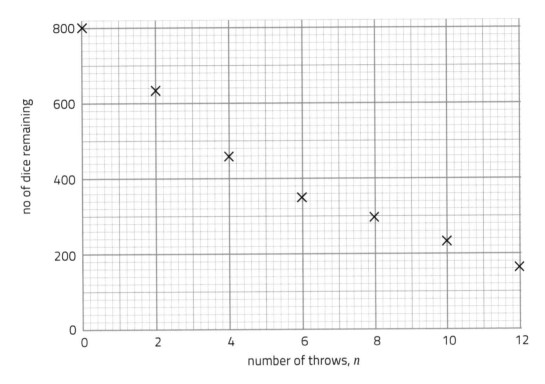

Comment on whether the results agree with the expected 'half-life' of the dice. [3]

....

....

....

....

....

(c) The students paint a second face black on each of the dice and repeat the experiment. Draw the expected decay graph on the grid above. [3]

Q8 β particles lose kinetic energy when they pass through the atoms of an absorber. They do this by interactions with the electrons in the atoms.

Their range in the absorber is the distance they travel until they lose all their kinetic energy.

The graph shows how the range of β particles **in water** depends upon their energy.

The range of β particles is inversely proportional to the density of the absorber.

β particle energy / MeV

(a) Determine the range of 1.0 MeV β particles in glass. [3]
$[\rho_{water} = 1.0 \times 10^3 \text{ kg m}^{-3}; \rho_{glass} = 2.5 \times 10^3 \text{ kg m}^{-3}]$

....

....

....

....

....

(b) Dylan claims the graph shows that β particles lose energy more quickly towards the end of their path than near the beginning. Evaluate whether this is correct. [2]

....

....

....

....

(c) Some Fire Exit signs are illuminated by fluorescent tubes, made of 1 mm thick glass. The tubes contain tritium, (^3_1H), a radioactive isotope of hydrogen, which emits β⁻ particles of energy 0.1 MeV. A 'phosphor' which coats the interior of the tubes, emits light when hit by β particles. The tubes do not need an electrical supply.

(i) Tritium is produced by the absorption of a neutron by lithium-6 (^6_3Li). Complete the equation for this reaction: [2]

$$^6_3Li \; + \; n \; \longrightarrow \; ^3_1H \; +$$

(ii) Write the decay equation for tritium: [2]

$$^3_1H \; \longrightarrow$$

(iii) Use the information above to explain why people in the vicinity of the tubes are in no danger from the β⁻ particles. [2]

....

....

....

Q9 Two technicians measured the count rate, C (in counts per second), over 1000 s, from a freshly produced radioactive sample. They plotted a graph of $\ln C$ against time.

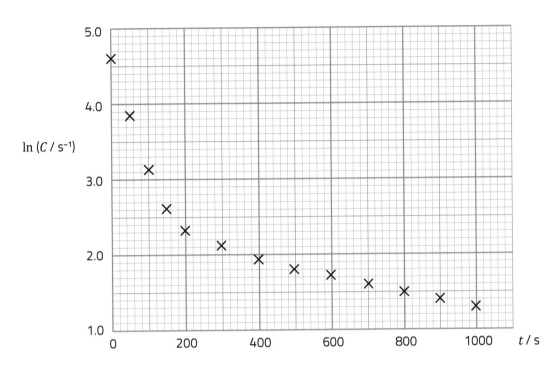

Dominic said that there must be two different radioisotopes in the sample and that after about 400 s, the one with the shorter half-life had decayed to a negligible amount.

(a) State how the results support Dominic's suggestion. [2]

....

....

....

(b) Use the results after 400 s to determine, for **isotope 2** (the one with the longer half-life):

(i) the decay constant. [2]

....

....

....

(ii) the count rate received at time $t = 0$. [2]

....

....

....

(c) Determine the initial count rate from **isotope 1** (the one with the shorter half-life). [2]

....

....

....

Q10 The radioactive nuclide, protoactinium-234, $^{234}_{91}Pa$, is sometimes used in schools for half-life determination because of its very short half-life. $^{234}_{91}Pa$ is produced by the decay of an isotope of thorium (Th), which is formed by the alpha decay of an isotope of uranium (U), which has a proton number of 92.

A website gives the half-life of ^{234}Pa as 1.17 minutes.

(a) Complete the decay equations to show the formation of protoactinium-234 from uranium: [3]

$$U \longrightarrow Th \quad +$$

$$Th \longrightarrow {}^{234}_{91}Pa \quad +$$

(b) In a classroom experiment, a student measured an initial count of 470 ± 22 over a period of 10 seconds. In a second reading 3.0 minutes later, the count was 86 ± 9. A preliminary measurement showed that background radiation was negligible compared to these readings.

Evaluate whether the student's results are consistent with the website data. [5]

....

....

....

....

....

....

....

(c) Describe how the student could obtain better data to compare with the website value of half-life. Briefly describe the method of analysis.

[Note that, with this apparatus, it is not possible to obtain an initial count rate greater than about 500 in 10 seconds.] [3]

....

....

....

....

....

Q11 A beam of β particles is passed through two narrow slits and is detected using a GM tube.

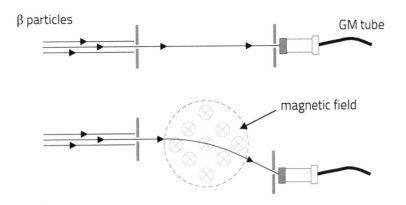

β particles

GM tube

magnetic field

When a magnetic field is applied between the slits, the second slit has to be moved downwards in order to detect the β particles.

(a) Explain this observation and deduce whether the β particles are β⁻ or β⁺. [4]

....

....

....

....

....

(b) Explain how the observations would be different if the beam consisted of α particles or γ photons. [3]

....

....

....

....

....

Question and mock answer analysis

Q&A 1

A group of students investigated two radioactive sources. They first measured the background count over a period of 1 minute using a radiation detector.

Result: Background count in 1 minute = 24.

(a) They set up the radiation detector 10 cm away from a beta (β) source and measured the count over 1 minute. The area of radiation detector was 1.0 cm².
Result: Count in 1 minute = 864.

Estimate the activity of the beta source, explaining your reasoning. [4]

(b) The students then investigated the absorption in lead of the radiation from a gamma source. They inserted a series of 0.50 cm thick pieces of lead between the source and the detector and recorded the corrected count, C, in 1 minute. Their results were as follows:

Absorber thickness, x / cm	C
0.00	572
0.50	308
1.00	233
1.50	144
2.00	81
2.50	67
3.00	40

The expected relationship is $C = C_o e^{-\mu x}$ where C_0 is the count due to source with no absorber, and μ is a constant.

(i) Show that a plot of $\ln C$ against x is expected to be a straight line. [2]

(ii) Use the grid to plot a graph of $\ln C$ against x and draw a suitable line. Error bars are not required. [8 cm × 12 cm grid supplied] [4]

(iii) Determine a value for μ and give an appropriate unit. [3]

(iv) The source and detector separation was a constant 10.0 cm. Explain, using an example, why the separation needed to be a controlled variable. [2]

What is being asked

Part (a) is a synoptic question from Unit 1 requiring application of the inverse square law of radiation. It also requires the handling of background corrections. This is an AO2 question. Part (b) is an experimental analysis based on a specified practical. The theory does not form part of the content of this topic; hence the expected relationship is given. Parts (i) and (ii) are standard AO2. Part (iii) is classed as AO3 because the method of using the results to obtain the answer is not flagged up. Part (iv) relates to the design of the experiment and is AO3.

Mark scheme

Question part			Description	AOs			Total	Skills	
				1	2	3		M	P
(a)			Subtraction of background [1] Conversion of time units [1] Application of $4\pi r^2$ [1] Activity = 17 kBq [1] Allow ecf on 864, using minutes and πr^2		4		4	1 1	4
(b)	(i)		Taking logs correctly, e.g. $\ln C = \ln C_0 - \mu x$ [1] Clear comparison with $y = mx + c$ [1]		2		2	2	

	(ii)	Linear axes labelled with unit on x axis and no unit on $\ln C$ axis [1] Scales chosen so points occupy at least 50% of each axis [1] Points plotted correctly within <1 square [1] Best fit straight line (by eye) drawn [1]		4		4			4
	(iii)	$\dfrac{\Delta y}{\Delta x}$ used for gradient (ignore sign) [1] Widely separated points on graph used. [1] $\mu = 0.86$ cm^{-1} **unit** [tolerance of ± 0.05] [1]			3	3	1		3
	(iv)	Inverse square law referred to or implied (accept: the count rate is lower at greater distances <u>because</u> radiation spread out) [1] Calculation, e.g. at 20 cm expect 143 counts with no absorber [1]			2	2	1		2
Total			0	10	5	15	6		13

Rhodri's answers

(a)   Corrected count = 842 ✓

   Fraction = $\dfrac{1.0 \text{ cm}^2}{\pi \times 10^2}$ = 3.18×10^{-3} ✗

   so 3.18×10^{-3} A = 842 ✗

   so A = 260 000 Bq ✓ ecf

> **MARKER NOTE**
> Rhodri corrected for background (first mark) but not for time and he incorrectly calculated the area of a 10 cm sphere.
> However, he obtained the last mark ecf.
>
> **2 marks**

(b)(i) $\log C = \log C_0 - \mu x$ ✓

   so a straight line ✗ (not enough)

> **MARKER NOTE**
> A clear comparison with the equation of a straight line needed for the second mark.
>
> **1 mark**

(ii)

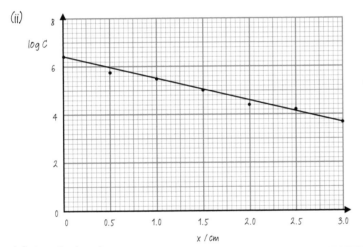

> ✓
> ✗
> ✓
> ✗
>
> **MARKER NOTE**
> Linear scales and correct labelling gives the first mark. Log is accepted for ln or log$_e$. The second mark is withheld because the points occupy less than half the vertical axis.
> Plots are correct – third mark. The line mark is not given because there are more points below the line than above – Rhodri has just joined the first and last points.
>
> **2 marks**

(iii) $\log C = \log C_0 - \mu x$.

   $\log C_0 = 6.35$; when $x = 2$, $\log C = 4.39$ ✓

   So $4.60 = 6.35 - 2\mu$

   $\mu = \dfrac{6.35 - 4.39}{2}$ ✓ $= 0.88$ (2 dp) ✗ (unit)

> **MARKER NOTE**
> The use of the equation is equivalent to using the gradient, as long as points from the line are used. The points are widely spaced. The unit is missing so the last mark is not given.
>
> **2 marks**

(iv) The radiation spreads out so the further away, the lower the counts ✓, so it wouldn't be a fair test. e.g. if it is moved to 15 cm the count would be lower. ✗ (not enough)

> **MARKER NOTE**
> The first mark is clearly given (radiation spreading out). A calculation is needed for the second mark.
>
> **1 mark**

> **Total**   **8 marks /15**

Ffion's answers

(a) At 10 cm the radiation spreads out to
area = $4\pi \times 10^2 = 1257$ cm^2 ✓
So activity = 1257×864 ✗
= 1.09×10^6 counts per min
= 18 kBq ✓✓ ecf

> **MARKER NOTE**
> The only mark that Ffion missed is the correction for background (subtracting 24). In fact the examiner would have given her the marks for the previous line because she gave a correct unit for activity there.
>
> **3 marks**

(b)(i) Equation of a straight line is $y = \mu x + c$.
If $C = C_0 e^{-\mu x}$, $\ln C = \ln C_0 - \mu x$ ✓
This has the same form as $y = mx + c$
if we plot $\ln C$ on the y axis.
The gradient is $-\mu$ ✓

> **MARKER NOTE**
> Ffion hits both marking points here and gains full marks.
>
> **2 marks**

(ii)

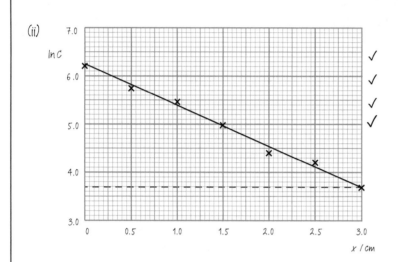

> **MARKER NOTE**
> Full marks: Ffion has chosen a ln C scale so that the points occupy more than half the axis.
> Her line has a gradient which reflects the points; the points are scattered equally above and below the line.
> The dotted line is for part (iii).
>
> **4 marks**

(iii) $\mu = \text{gradient} = \dfrac{6.25 - 3.69}{3.0 \text{ cm}}$ ✓✓ (bod)
= 0.85 cm^{-1} ✓

> **MARKER NOTE**
> Ffion has made two mistakes of sign in what she writes. As she says in (b)(i), the gradient is $-\mu$ and her expression is minus the gradient. However, these are not penalised. Her unit is correct.
>
> **3 marks**

(iv) If the distance increases, the count rate will decrease because of the inverse square law ✓
e.g. if it is doubled, the count will become $\frac{1}{4}$.
So to make a comparison the distance needs to be made the same. ✗ (not enough)

> **MARKER NOTE**
> The second mark is a difficult one to achieve. An answer such as, "20 cm would give a count of only 77 counts with 0.5 cm of lead – but not because of absorption," would get the mark.
>
> **1 mark**

> **Total** **13 marks /15**

Section 6: Nuclear energy

Topic summary

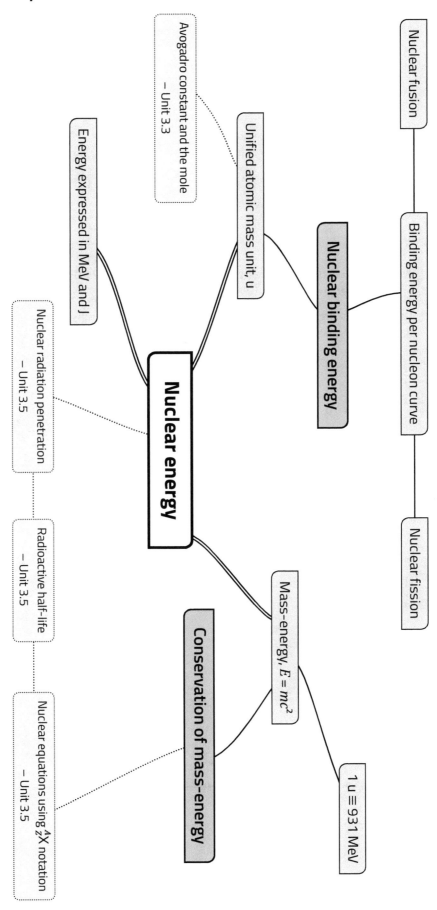

Q1 When asked to define the *binding energy* of a nucleus, Julia wrote:

This is the energy needed to hold the particles of a nucleus together.

Octavia said that this could not be right because that would mean that a nucleus would have a greater mass than the protons and neutrons in it.

(a) Explain Octavia's comment. [2]

....

....

....

(b) Write a correct definition of nuclear binding energy. [2]

....

....

....

Q2 A website gives the mass of a neutral ^1_1H atom as 1.007 825 032 u and the ionisation energy (the energy needed to remove the electron) as 13.6 eV.

Use the following data to evaluate whether the sum of the masses of a proton and an electron is different from the atomic mass to this number of significant figures:

$u = 1.66 \times 10^{-27}$ kg $e = 1.60 \times 10^{-19}$ C $c = 3.00 \times 10^8$ m s^{-1} [3]

....

....

....

....

....

Q3 A data book gives the following mass data in u:

electron: 0.000 549   proton: 1.007 276   neutron: 1.008 665 ^4_2He atom: 4.002 604

Use this information to calculate:

(a) the binding energy of a ^4_2He atom [3]

....

....

....

....

(b) the binding energy per nucleon of a ^4_2He atom [1]

....

Q4 The intensity of electromagnetic radiation from the Sun incident on the Earth's atmosphere is 1370 W m^{-2}. The radius of the Earth's orbit is 1.50×10^{11} m.

(a) Calculate the power emitted by the Sun as electromagnetic radiation. [2]

....

....

....

(b) It is commonly stated that the Sun loses 4 million tonnes of mass each second due to the output of radiation. Evaluate this statement. [1 tonne = 1000 kg] [2]

....

....

....

Q5 A science data book gives the following data for uranium-235 which decays by α emission:

Atomic mass = 235.043 930 u; density = 18.8×10^3 kg m^{-3}; half-life = 7.1×10^8 years.

(a) Calculate the activity of 1.0 kg of pure $^{235}_{92}U$. [4]

....

....

....

....

....

(b) Michael and Jonathan were discussing how a 1.0 kg lump of $^{235}_{92}U$ would feel. They agreed that it might be warm to the touch. Michael suggested you wouldn't notice this.

(i) Explain why it might be warm. [2]

....

....

....

(ii) Discuss Michael's suggestion. [4]

Additional data: $m(^{231}_{90}Th) = 231.036\ 304$ u; $m(^4_2He) = 4.002\ 604$ u

....

....

....

....

....

Q6 Tritium, ^3_1H, is a radioactive isotope of hydrogen. It is hoped to use tritium in a nuclear fusion reactor.

Tritium is made by bombarding the lithium isotope ^6_3Li with neutrons in a specially designed fission reactor. The reaction also produces one other nuclide in the reaction.

The grid shows the binding energy per nucleon of various nuclides given to the nearest 0.1 MeV nuc^{-1}.

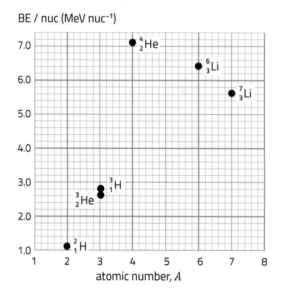

(a) Complete the equation of the nuclear reaction between ^6_3Li and a neutron which produces tritium:

$^6_3Li + ^1_0n \longrightarrow$ [2]

(b) Tritium decays by α, β^- or β^+ emission into one of the other nuclides identified on the chart.

Write the decay equation for tritium:

$^3_1H \longrightarrow$ [2]

(c) In the proposed reaction in the fusion reactor, tritium reacts with deuterium, another isotope of hydrogen, ^2_1H, to produce ^4_2He and one other particle.

Determine the energy released in the fusion of 1.0 kg of an appropriate mixture of tritium and deuterium. [5]

....

....

....

....

....

....

....

....

Q7 The Sun, a middle-aged star, derives its energy from the fusion of hydrogen to helium in a process which can be summarised as:

$$4\,^{1}_{1}H \longrightarrow \,^{4}_{2}He + 2\,^{0}_{-1}e$$

where $^{1}_{1}H$ and $^{4}_{2}He$ are the atomic symbols. This stage in the life-cycle of the Sun is expected to last 9×10^{9} years in total.

The Sun will later go through a stage of 'helium-burning' in which the helium produced by hydrogen fusion reacts to form carbon in the so-called triple-alpha process:

$$^{4}_{2}He + \,^{4}_{2}He + \,^{4}_{2}He \longrightarrow \,^{12}_{6}C$$

Data: $m(^{1}_{1}H) = 1.007\,825$ u; $m(^{4}_{2}He) = 4.002\,604$ u; $m(^{12}_{6}C) = 12$ u (exactly); $m(^{0}_{-1}e) = 0.000\,549$ u

(a) Show that the fusion of four $^{1}_{1}H$ atoms to $^{4}_{2}He$ yields 25.7 MeV. [2]

(b) Calculate the energy released in the triple-alpha process. [2]

(c) During the 'helium-burning' phase the Sun will have 10× its current diameter; its surface temperature will be 90% its current kelvin value.

(i) Show that the luminosity of the Sun will be approximately 65× its current value. [2]

(ii) Use the information at the start of the question and your answers to estimate the length of time that the helium-burning phase is able to last. Show your reasoning. [4]

Q8 Elements with nucleon numbers greater than 56 are produced in supernova explosions and in the merger of neutron stars. In these very dense conditions with large numbers of free neutrons, nuclei such as iron-56 ($^{56}_{26}Fe$) absorb neutrons and then undergo β^- decay in a process which builds up heavier nuclei.

The production of $^{235}_{92}U$ from $^{56}_{26}Fe$ can be summarised by:

$$^{56}_{26}Fe \ + \ \text{............} \ ^{1}_{0}n \ \longrightarrow \ ^{235}_{92}U \ + \ \text{............} \ ^{0}_{-1}e \ + \ \text{............} \ ^{0}_{0}\overline{\nu}_e$$

Complete the equation with the correct numbers of particles and explain your reasoning. [4]

....

....

....

....

....

....

Q9 The mass of a $^{4}_{2}He$ atom is 4.002 604 u. The mass of a $^{8}_{4}Be$ atom is 8.005 305 u. $^{8}_{4}Be$ decays (with a very short half-life of 10^{-16} s) into two $^{4}_{2}He$ atoms.

(a) Explain why $^{8}_{4}Be$ is able to decay to $^{4}_{2}He$ in this way and suggest why the decay has such a short half-life. [2]

....

....

....

(b) If the $^{8}_{4}Be$ is stationary, the two resulting $^{4}_{2}He$ nuclei are seen to move off in opposite directions.

(i) Explain this observation. [2]

....

....

....

(ii) Calculate the speeds of the $^{4}_{2}He$ nuclei. [The mass of electrons can be ignored.] [4]

....

....

....

....

....

....

Question and mock answer analysis

Q&A 1

In a prototype nuclear fusion reactor, a ^3_1H nucleus and a ^2_1H nucleus collide head-on at high speed. They combine to produce a stationary unstable nucleus A, which decays with a half-life of 7.6×10^{-22} s into a nucleus of ^4_2He and another particle B. This is illustrated in the following diagram:

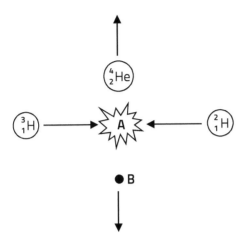

(a) Identify the unstable nucleus A and particle B, briefly explaining your answer. [1]

(b) Identify the nuclear interaction responsible for the decay of A **giving two reasons** for your answer. [2]

(c) Explain why the two hydrogen nuclei need to approach at high speed for these reactions to occur. [2]

(d) Alex and Eirian discussed these reactions in terms of mass-energy. Alex claimed the total mass of the ^3_1H and ^2_1H must be more than the mass of A. Eirian disagreed and said that by conservation of mass-energy, these masses must be equal.
Discuss who, if either, is correct. [2]

(e) The masses of some of the particles are given:

$m(A) = 5.010\ 959$ u; $m(^4_2He) = 4.001\ 505$; $m(B) = 1.008\ 665$ u

(i) Calculate the energy released in the decay of A. Give your answer in J. [3]

(ii) Explain why the speed of particle B is 4.0 times that of the ^4_2He nucleus. [2]

(iii) Use (i) and (ii) to determine the speed of the ^4_2He nucleus. [3]

(iv) In order for the fusion of the hydrogen nuclei to occur, they have to approach within 1.0×10^{-14} m. By considering energy conservation, calculate the total kinetic energy of these two particles before the collision and hence estimate the necessary temperature of the gas in the reaction chamber. [5]

What is being asked?

The question involves two reactions, nuclear fusion and an unusual radioactive decay. It brings many areas of the A level physics specification into play. Section 3.6 relies heavily on the knowledge of other sections of the specification, so many questions are synoptic in nature. Part (a) relies on the concepts of Section 3.5. Particle physics (1.7) is used in parts (b) and (c). Part (d) is an AO3 question which is purely Section 3.6. Part (e) brings in momentum conservation in (ii), kinetic energy in (iii). The last part, (e)(iv), another AO3 question, is another synoptic part involving electric fields (4.2) and kinetic theory (3.3).

Mark scheme

Question part			Description	AOs			Total	Skills	
				1	2	3		M	P
(a)			A = $^5_2$He (nucleus), **and** B = neutron Correct reasoning including proton and neutron numbers in $^5_2$He (or equiv) [1]		1		1		
(b)			Strong interaction / force because: • Half-life {so short / typical of strong} [1] • Only hadrons/quarks involved **or** no change in quark flavour [1]			2	2		
(c)			Nuclei [both] positively charged so repel [1] Need to approach closely for fusion to occur [1]	2			2		
(d)			The initial <u>kinetic energy</u> (of the $^3_1$H and $^2_1$H) also has mass. [1] (Mass/energy is conserved) so the mass (of A) is more than the total mass of $^3_1$H and $^2_1$H; hence neither is correct. [1]			2	2		
(e)	(i)		Mass loss = 7.89×10^{-4} (u) [1] [no sign penalty] Energy released = 0.735 MeV [1] $= 1.176 \times 10^{-13}$ J [1]			3	3	3	
	(ii)		The momenta of $^4_2$He and {B/neutron} are equal (and opposite) [1] $m(^4_2$He$) = 4.0 \times m$(B) [1]			2	2		
	(iii)		If speed of $^4_2$He = v, KE $= \frac{1}{2}(4.00v^2 + 1.01 \times (4.0v)^2)[\times 1.66 \times 10^{-27}]$ J [1] $= 1.67 \times 10^{-26}v^2$ ∴ $1.67 \times 10^{-26}v^2 = 1.176 \times 10^{-13}$ J [1] ∴ Speed of $^4_2$He $= 2.65 \times 10^6$ m s^{-1} [1]			3	3	2	
	(iv)		Use of $\frac{1}{4\pi\varepsilon_0}\frac{Q_1 Q_2}{r}$ [1] [e.g. 2.30×10^{-14} J seen] Use of conservation of energy [1] Use of $(\frac{3}{2})kT$ with mean particle energy [i.e. 1.15×10^{-14} J ecf] [1] ~800 MK [or 560 MK] obtained [1] ecf Suggestion that a significantly lower temperature is will produce fusion. [1]			5	5	3	
Total				2	9	9	20	8	

Rhodri's answers

(a) It must be $^5_2$He because it has the right number of protons (2) and neutrons (3). ✗

MARKER NOTE
B not identified as a neutron.
0 marks

(b) It must be a strong interaction because it happens in such a short time – 10^{-22} s. ✓
Second reason ???

MARKER NOTE
Only one of the two reasons given.
1 mark

(c) The strong force has a very short range so the particles have to collide in order for the reaction to happen, which needs a high energy ✓ bod .

MARKER NOTE
Rhodri has given the information about the distance of approach needed. He hasn't related this to overcoming the repulsion. This might be hinted at with the need for a high energy, but that is not clearly stated.
1 mark

(d) The total mass-energy is the same for all three stages of the process as long as we remember to include the kinetic energy of the particles (because there is nowhere else for the energy to go).✓ So the $^3_1$H and $^2_1$H masses will be more than when they are at rest and Eirian is correct. ✓

MARKER NOTE
The examiner has used judgement here to award marks for a different interpretation. Rhodri considers the masses of the $^3_1$H and $^2_1$H to include that of the kinetic energy and in that sense, Eirian is correct.
2 marks

(e)(i) Mass of products = 5.01017 u
Mass loss = 0.000789 u ✓
= 1.31×10^{-30} kg
∴ Energy released = mc^2
= 1.18×10^{-13} J ✓✓

MARKER NOTE
Rather than use the conversion factor 931 MeV/u, Rhodri has converted to kg and used $E = mc^2$. This is perfectly acceptable.
3 marks

(ii) The momentum of B is equal and opposite to the momentum of the $^4_2$He ✓. B has $\frac{1}{4}$ of the mass so 4× the speed. ✓

MARKER NOTE
Neatly expressed.
2 marks

(iii) Ratio of masses 1:4. Using $\frac{m}{m \times M}$, $^4_2$He gets $\frac{1}{5}$ of the total energy = 2.36×10^{-14} J ✓
So $\frac{1}{2}$ 4.00 × 1.66 × 10^{-27} v^2 = 2.36×10^{-14} ✓
∴ $v = 1.86 \times 10^6$ m s^{-1} ✗

MARKER NOTE
Rhodri has not used the way in from (ii) but has got off to a good start using ratios and worked out the energy of the ^{4}He nucleus. Unfortunately he has forgotten the ½ in calculating v, so loses the third mark.
2 marks

(iv) Energy released = 1.18×10^{-13} J ✗
This is shared between the particles so mean energy = 5.9×10^{-14} J
$E = \frac{3}{2}kT$ ∴ $T = \frac{2 \times 5.9 \times 10^{-14}}{3 \times 1.38 \times 10^{-23}}$ ✓ ecf
= 2.85×10^9 K ✓ecf
Even at lower temperatures (e.g. $\frac{1}{2}$ this) some molecules have enough energy ✓

MARKER NOTE
Rhodri has not understood this synoptic question which requires knowledge of Unit 4. He has used the KE of the daughter particles and hence loses the first two marks. However, his analysis of the relationship between energy and temperature, which are Unit 3 concepts, is spot on.
3 marks

Total **14 marks /20**

Ffion's answers

(a) $^3_1H + ^2_1H \longrightarrow ^5_2He\ (A) \longrightarrow ^4_2He + ^1_0n\ (B)$

This balances so ^5_2He and a neutron ✓

MARKER NOTE
Not what was expected but Ffion clearly knows what the reactions are, and the explanation implies consistent number of protons and neutrons.
1 mark

(b) Only protons and neutrons are involved ✓ (no leptons or photons) and it happens so quickly ✓, so it must be the strong interaction.

MARKER NOTE
'Only protons and neutrons' is as good here as 'only baryons'.
2 marks

(c) The ^3_2He and ^2_1He have got to get very close for fusion.✓ They are both positively charged so repel each other.✓ If they had low speeds, they wouldn't get close enough.

MARKER NOTE
Both marking points are clearly made. The last sentence was not required on this occasion.
2 marks

(d) Producing A is like a reverse fission reaction. In a fission reaction, there is a mass loss which results in the energy output. ✗ [Not enough]

Hence the mass of the $^3_2He + ^2_1He$ must be less than A and they are both wrong. ✓

MARKER NOTE
The first marking point is not quite there. The <u>kinetic</u> energy of the hydrogen nuclei needs to be clearly linked to mass-energy.
The second point is clearly made.
1 mark

(e) (i) $\Delta m = 5.010\ 959 - 4.001\ 505 - 1.008\ 665$ ✓
$\qquad = 0.000\ 789\ u$ ✓

$1\ u = 931\ MeV$
So energy released $= 0.000\ 789 \times 931$
$\qquad\qquad\qquad = 0.735\ MeV$ ✓

MARKER NOTE
Strictly Δm represents the mass gain rather than the mass loss but no penalty is applied on this occasion. Ffion gave the answer in MeV which is accepted.
3 marks

(ii) The momentum of B is equal and opposite to the momentum of the ^4_2He ✓. B has $\frac{1}{4}$ of the mass so $4\times$ the speed. ✓

MARKER NOTE
Neatly expressed.
2 marks

(iii) If speed of ^4_2He is v its KE is:
$\frac{1}{2}\ 4.00 \times 1.66 \times 10^{-27}\ v^2 = 3.32 \times 10^{-27}v^2$
and KE of neutron $= 13.28 \times 10^{-27}v^2$
$\therefore$ Adding: $1.66 \times 10^{-26}v^2$ ✓ $= 1.18 \times 10^{-13}$ ✓
$\therefore \longrightarrow v = 2.67 \times 10^6\ m\ s^{-1}$ ✓

MARKER NOTE
A very clear answer which ticks all the boxes. Note that Ffion had to do the MeV to J conversion which she didn't do in (e)(i).
3 marks

(iv) $PE = 9 \times 10^9 \times \dfrac{(1.6 \times 10^{-19})^2}{1.0 \times 10^{-14}} = 2.30 \times 10^{-14}\ J$✓

$\therefore$ Initial KE of particles $= 2.30 \times 10^{-14}\ J$ ✓

$\therefore$ Using $KE = \frac{3}{2}kT \longrightarrow T = 1.11 \times 10^9\ K$ ✓ecf

MARKER NOTE
Ffion used the potential energy formula correctly and (by implication) energy conservation for the first two marks. Her calculation of temperature was not correct because she didn't use the mean kinetic energy, so she lost the third mark. The difficult last mark required a comment taking into account energy distribution.
3 marks

Total **17 marks /20**

Unit 4: Fields and Options

Section 1: Capacitance

Topic summary

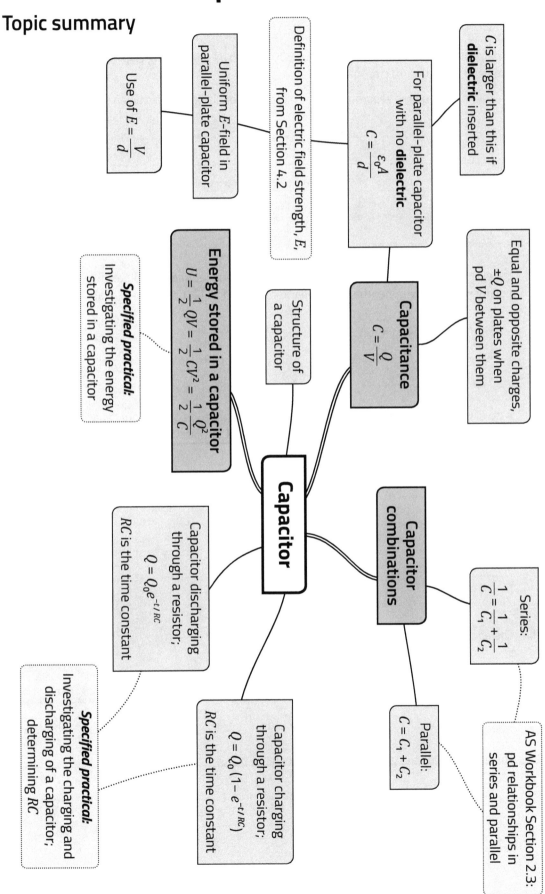

Capacitor

Structure of a capacitor

Capacitance
$$C = \frac{Q}{V}$$

Equal and opposite charges, $\pm Q$ on plates when pd V between them

For parallel-plate capacitor with no **dielectric**
$$C = \frac{\varepsilon_0 A}{d}$$

C is larger than this if **dielectric** inserted

Definition of electric field strength, E, from Section 4.2

Uniform E-field in parallel-plate capacitor

Use of $E = \frac{V}{d}$

Energy stored in a capacitor
$$U = \frac{1}{2}QV = \frac{1}{2}CV^2 = \frac{1}{2}\frac{Q^2}{C}$$

Specified practical: Investigating the energy stored in a capacitor

Capacitor discharging through a resistor;
$$Q = Q_0 e^{-t/RC}$$
RC is the time constant

Capacitor charging through a resistor;
$$Q = Q_0(1 - e^{-t/RC})$$
RC is the time constant

Specified practical: Investigating the charging and discharging of a capacitor; determining RC

Capacitor combinations

Series:
$$\frac{1}{C} = \frac{1}{C_1} + \frac{1}{C_2}$$

Parallel:
$$C = C_1 + C_2$$

AS Workbook Section 2.3: pd relationships in series and parallel

Q1 Determine the charges on each plate of a 22 mF capacitor when a pd of 12 V is placed between them.

[2]

....

....

....

Q2 A parallel-plate capacitor of capacitance 500 pF is to be made using two flat metal plates, each measuring 10 cm × 10 cm.

(a) Calculate the required separation of the plates, if there is only air between them. [3]

....

....

....

....

....

(b) Ludovic intends to make the capacitor, but with the gap between the plates filled with an insulating polymer. Discuss whether the plate separation should be that calculated in (a). [2]

....

....

....

Q3 A pd of 30 V is applied between the plates of an air-spaced parallel-plate capacitor. Each plate has an area of 64 cm², and the plate separation is 0.40 mm. Calculate:

(a) The charge on either plate. [3]

....

....

....

....

(b) The energy stored in the capacitor. [2]

....

....

....

(c) The electric field strength in the gap between the plates. [1]

....

....

....

Q4 A battery is connected across a parallel-plate capacitor, and then removed, leaving charges on the plates as shown:

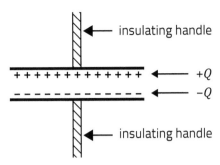

The plates are then pulled further apart, doubling their separation.

(a) Determine by what factor the energy stored in the capacitor is increased, stating clearly the principle on which your answer is based. [3]

....

....

....

....

....

(b) State where the extra energy has come from. [1]

....

Q5 Show that, for a given electric field strength, E, between the plates of a parallel-plate capacitor, the energy stored is proportional to the *volume* of the space between the plates. [2]

....

....

....

....

Q6 Write the *charges* (sign and magnitude) on the capacitor plates in the boxes provided. The space below the diagram is for your working. [4]

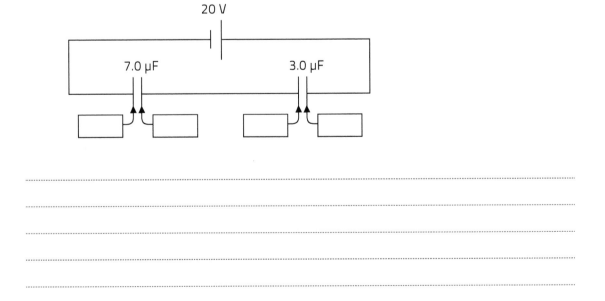

....

....

....

....

....

Unit 4 Practice questions

Q7 A capacitor combination is shown:

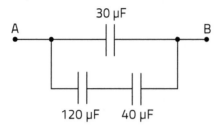

(a) Calculate its capacitance. [3]

....

....

....

....

(b) A battery of emf 12 V is connected across AB. Calculate:

(i) The charge that flows through the battery while the capacitors charge. [1]

....

(ii) The final pd across the 120 μF capacitor, giving your reasoning. [2]

....

....

....

Q8 A battery is connected across a 50 mF capacitor, C_1. The battery is then disconnected, leaving a pd of 9.0 V between the plates of C_1. An (initially) uncharged 50 mF capacitor, C_2, is now connected across C_1.

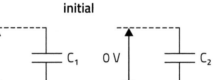

(a) Explain why the 'final' pd, V, is 4.5 V. [3]

....

....

....

....

(b) Calculate the change in total energy stored when C_2 is connected across C_1. [3]

....

....

....

....

Q9 In the circuit shown, the capacitor is initially uncharged. The switch is closed at time $t = 0$.

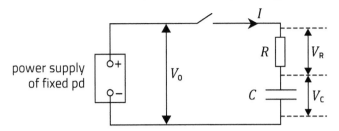

After time $t = 0$ the following equation applies:

$$Q = Q_0(1 - e^{-t/RC})$$

(a) What is the meaning of Q_0? [1]

....

(b) In each of the following parts, use the preceding equation to show that:

(i) $\qquad\qquad\qquad\qquad V_C = V_0(1 - e^{-t/RC})$ [2]

....

....

....

(ii) $\qquad\qquad\qquad\qquad V_R = V_0 e^{-t/RC}$ [2]

....

....

....

(iii) $\qquad\qquad\qquad\qquad I = I_0 e^{-t/RC}$ [2]

....

....

....

(c) (i) Starting from definitions of capacitance and resistance, show that the SI unit of Q_0/RC is the ampère. [3]

....

....

....

....

(ii) Explain why the gradient of the tangent at $t = 0$ to a graph of Q against t is Q_0/RC. [2]

....

....

....

Q10 The pd, V_c, between the plates of a capacitor is plotted against time as the capacitor discharges through a resistor.

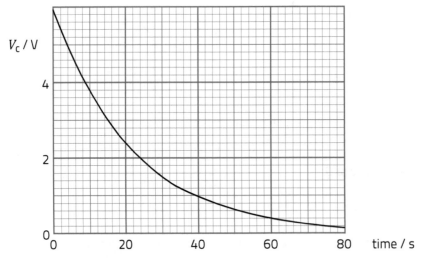

(a) Draw a circuit diagram of the arrangement that might have been used to obtain the results. Include a means of charging the capacitor beforehand. [3]

(b) (i) Using three points on the graph, verify that V_c decays exponentially with time. [3]

....

....

....

....

....

(ii) Determine the *time constant* of the decay. [2]

....

....

....

(c) For a graph of ln (V_c/volt) against time, give the values of the gradient and the intercept on the ln (V_c/volt) axis. [2]

....

....

....

Question and mock answer analysis

Q&A 1

Lauren investigated the energy stored in a capacitor using the apparatus shown:

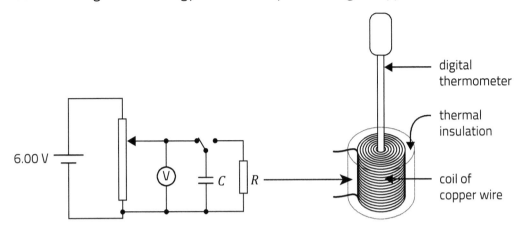

Lauren read the temperature, θ_1, of the coil of copper wire. She then charged the capacitor to a pd, V, and connected it across the coil (R in the circuit diagram). She read the coil's temperature, θ_2, after the capacitor had been discharging for 30 s. She repeated the procedure for six more pds, up to 5.0 V, and then again for all seven pds. Her plot of $(\theta_2 - \theta_1)$ against V^2 is given.

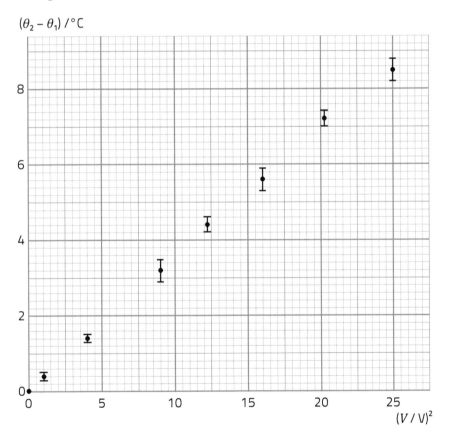

(a) Name the part of the circuit that enabled Lauren to select the various pds. [1]

(b) Suggest how Lauren decided on the vertical position of her points, and the error bars that she used. [2]

(c) Lauren's decision to plot $(\theta_2 - \theta_1)$ against V^2 was based on the equation:

$$\tfrac{1}{2} CV^2 = mc\,(\theta_2 - \theta_1)$$

in which C = capacitance of capacitor = 5.00 F

m = mass of copper in coil = 15.8 g

c = specific heat capacity of copper = 385 J kg^{-1}°C^{-1}

(i)   State what quantities are represented by $\frac{1}{2}CV^2$ and $mc\,(\theta_2 - \theta_1)$. [1]

(ii) Evaluate to what extent Lauren's data, as plotted, support the equation. [7]

(d) (i)   Decide, using a suitable calculation, whether or not 30 s was a long enough time to allow for capacitor discharge. [Resistance of coil = 2.0 Ω] [3]

(ii) Suggest why allowing a much longer time (for example 5 minutes) before reading the temperature, θ_2, would be likely to cause error. [1]

What is being asked

This question is centred on a specified practical investigation of the energy stored in a capacitor. It is not assumed that you will be familiar with exactly the same apparatus as that described, but enough information is given for you to figure out how it works. The question (like the investigation itself) is synoptic, as electric circuits and thermal physics are both touched upon.

Part (a) is a straightforward interpretation of the diagram, AO1; part (b) tests your familiarity with using experimental data, applied to these results, hence AO2. Part (c)(i), AO1, tests whether you understand the physical principle on which the equation is based, essentially recognition of two energy terms. In (c)(ii) you are left to your own devices to decide on your strategy, which makes it AO3, calling for calculations and conclusions. In (d), part (i) is another evaluation, AO3, with more than one way of answering it for three marks, but the single mark in part (ii) suggests there is really only one point to be made.

Mark scheme

Question part		Description	AOs			Total	Skills	
			1	2	3		M	P
(a)		Potential divider. Accept 'potentiometer' [1]	1			1		1
(b)		Point plotted at mean of the two temp readings [1]		2		2		2
		Error bar runs between the two temperature readings [1]						
(c)	(i)	Energy [initially] in capacitor; gain in internal energy [accept thermal energy, heat, random energy] of coil [finally]	1			1		
	(ii)	At least one straight line through origin drawn, passing through all error bars, or all except the one furthest from the origin [1]			7	7	3	6
		Comment made that data points fit straight line through origin, as equation predicts [1]						
		but the point furthest from origin is anomalous [1]						
		Gradient of any straight line (e.g. attempt at best fit, or maximum gradient) correctly calculated for the chosen line [1]						
		Maximum gradient stated to be = 0.37 or 0.35 [°C V^{-2}] [1]						
		Theoretical gradient = 0.41 [°C V^{-2}] [1]						
		Comment that even maximum gradient is too small, so [to this extent] data don't fit equation [1]						
(d)	(i)	Time constant [= 5.0 F × 2.0 Ω] = 10 s [1]			3	3	1	3
		Either						
		[e^{-3}=] 5.0% of original charge (or voltage/pd) or 2.5% of energy left after 30 s [1]						
		so long enough **or** but not long enough [1]						
		Or						
		30 s is [considerably] greater than time constant therefore long enough **or** but not long enough [1]						
		[2 marks maximum for this approach]						
	(ii)	Heat will escape through the thermal insulation [in such a long time] [1]		1		1		1
Total			2	3	10	15	4	13

Rhodri's answers

(a)   A variable resistor ✗

(b)   Point is plotted midway between the two temperatures measured for that voltage. ✓
Error bar represents the uncertainty. ✗ [Not enough]

(c) (i)   The left-hand side is the energy stored in the capacitor. The right-hand side is the heat given to the copper wire coil. ✓

(ii)

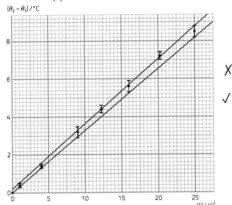

$(\theta_2 - \theta_1) / °C$

$(V / V)^2$

Points almost fit a straight line through the origin. This is right because equation says that $(\theta_2 - \theta_1)$ proportional to V^2. ✓

Least gradient $= \dfrac{8.2}{25} = 0.33\,°C\,V^2$ ✓

Largest $= (8.8 - 0.2)/25 = 0.34\,°C\,V^2$

In theory, gradient $= \dfrac{(C/2)}{mc}$

$= 2.5 \div 6.08$

$= 0.41\,°C\,V^2$. ✓

(d) (i) $CR = 5 × 2 = 10\,s$ ✓

30 s is 3 times as long, so the capacitor will be well discharged. ✓

(ii)   Perhaps the coil will have started to cool off! ✓

Ffion's answers

(a)   Variable potential divider ✓

(b)   Ends of error bar are at the two temperatures. ✓
Point is plotted at their mean. ✓

(c) (i) $CV^2/2$ is the energy in the charged capacitor.
$mc(\theta_2 - \theta_1)$ is the internal energy in the
copper after the discharge. ✗

(ii)

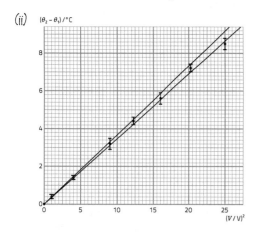

Rearranging equation:

$(\theta_2 - \theta_1) = \dfrac{C}{2mc} V^2$ so we should have a straight
line through the origin, ✓ of gradient

$\dfrac{C}{2mc} = \dfrac{5.0}{(2 = 0.0158 = 385)} = 0.41$ ✓

The points, all except the last one ✓, do seem to
fit a straight line through the origin ✓. Ignoring
the last point,

Lowest gradient $= \dfrac{8.7}{25} = 0.348$ ✓

Highest gradient $= \dfrac{9.2}{25} = 0.375$ ✓

So even the steepest line has too small a gradient,
so the fit is not so good after all. ✓

(d) (i) $Q = Q_0 e^{-t/RC}$ so $Q = Q_0 e^{-30/10} =$ ✓ 0.050

Therefore $\dfrac{V}{V_0} = \dfrac{1}{20}$ and $\left(\dfrac{V}{V_0}\right)^2 = \dfrac{1}{400}$, ✓
so almost no energy left in the capacitor. ✓

(ii)   The coil will have had time to lose heat
to the surroundings, so the temperature
rise will be lower than predicted. ✓

Total   **14 marks /15**

Section 2: Electrostatic and gravitational fields of force

Topic summary

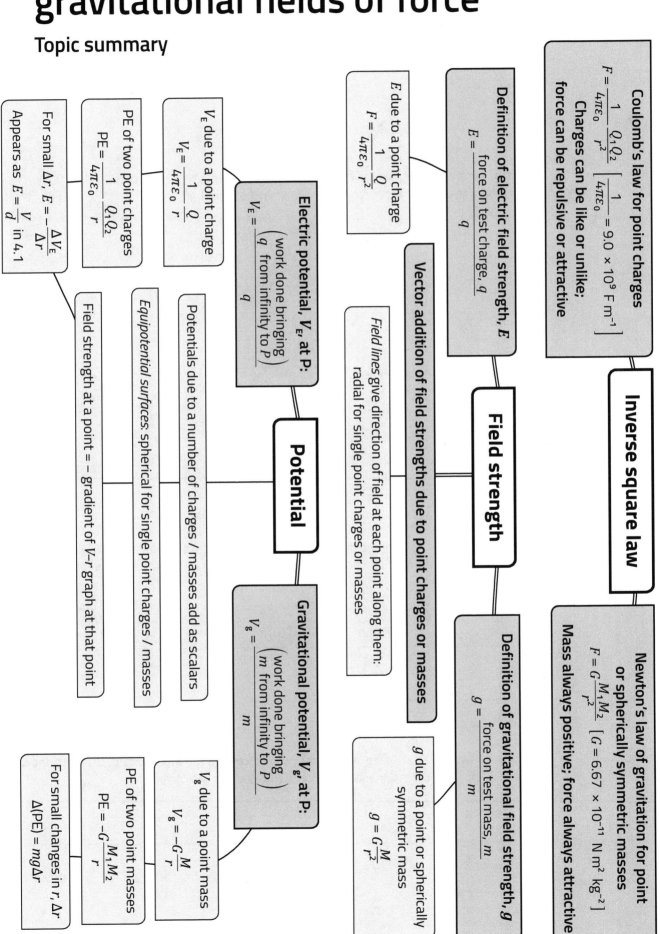

Coulomb's law for point charges

$$F = \frac{1}{4\pi\varepsilon_0}\frac{Q_1 Q_2}{r^2} \left[\frac{1}{4\pi\varepsilon_0} = 9.0 \times 10^9 \text{ F m}^{-1}\right]$$

Charges can be like or unlike; force can be repulsive or attractive

Inverse square law

Newton's law of gravitation for point or spherically symmetric masses

$$F = G\frac{M_1 M_2}{r^2} \quad [G = 6.67 \times 10^{-11} \text{ N m}^2 \text{ kg}^{-2}]$$

Mass always positive; force always attractive

Definition of electric field strength, E

$$E = \frac{\text{force on test charge, } q}{q}$$

E due to a point charge

$$F = \frac{1}{4\pi\varepsilon_0}\frac{Q}{r^2}$$

Field strength

Vector addition of field strengths due to point charges or masses

Field lines give direction of field at each point along them: radial for single point charges or masses

Definition of gravitational field strength, g

$$g = \frac{\text{force on test mass, } m}{m}$$

g due to a point or spherically symmetric mass

$$g = G\frac{M}{r^2}$$

Electric potential, V_E, at P:

$$V_E = \frac{\left(\begin{array}{c}\text{work done bringing} \\ q \text{ from infinity to } P\end{array}\right)}{q}$$

V_E due to a point charge

$$V_E = \frac{1}{4\pi\varepsilon_0}\frac{Q}{r}$$

PE of two point charges

$$PE = \frac{1}{4\pi\varepsilon_0}\frac{Q_1 Q_2}{r}$$

For small Δr, $E = -\frac{\Delta V_E}{\Delta r}$
Appears as $E = \frac{V}{d}$ in 4.1

Potential

Field strength at a point = − gradient of V–r graph at that point

Equipotential surfaces: spherical for single point charges / masses

Potentials due to a number of charges / masses add as scalars

Gravitational potential, V_g, at P:

$$V_g = \frac{\left(\begin{array}{c}\text{work done bringing} \\ m \text{ from infinity to } P\end{array}\right)}{m}$$

V_g due to a point mass

$$V_g = -G\frac{M}{r}$$

PE of two point masses

$$PE = -G\frac{M_1 M_2}{r}$$

For small changes in r, Δr
$$\Delta(PE) = mg\Delta r$$

Q1 Two identical small spheres, each of mass 2.00×10^{-7} kg, carry equal positive charges and hang on insulating threads from a fixed point. When in equilibrium they hang as shown.

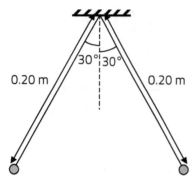

30° 30°

0.20 m   0.20 m

(a) Show that the electrostatic repulsive force on each sphere must be approximately 1.1×10^{-6} N. Use the space to the right above for a vector diagram if required. [3]

....

....

....

....

....

(b) Calculate the charge on each sphere. [3]

....

....

....

....

....

Q2 (a) The Moon's radius is 1737 km, and the gravitational field strength at its surface is 1.62 N kg^{-1} towards its centre. Show that the Moon's mass is approximately 7×10^{22} kg, stating an assumption that you make. [3]

....

....

....

....

(b) The Earth's mass is 5.97×10^{24} kg, and the mean distance between the centres of the Earth and Moon is 3.84×10^8 m. Calculate the gravitational pull of the Moon on the Earth. [2]

....

....

....

....

Q3 (a) The mass of a proton is 1.67×10^{-27} kg. For two protons a given distance apart, calculate the ratio:

$$\frac{\text{electrostatic force between protons}}{\text{gravitational force between protons}}$$

[3]

....

....

....

....

....

(b) Protons make up a sizeable fraction (by mass) of both the Sun and the Earth. Explain why the gravitational force between these bodies is far greater than the electrostatic force. [3]

....

....

....

....

....

Q4 (a) In the space on the left below draw eight field lines showing the electric field around an isolated negative charge. [2]

Diagram for part (a)
(to be completed)

Diagram for part (b)

(b) Some field lines in the area around equal and opposite point charges are shown on the right. By considering the fields due to individual charges, explain why the field line at P is in the direction shown. [3]

....

....

....

....

....

Unit 4 Practice questions

Q5   Equal and opposite charges are separated by a distance of 0.12 m, as shown:

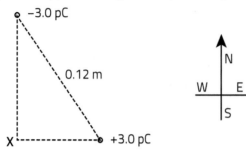

(a) (i)   Show clearly that the electric field strength at **X** due to the negative charge is 2.8 N C^{-1}
           Northwards.   [3]

_____

_____

_____

_____

(ii) Determine the electric field strength at **X** due to the positive charge.   [2]

_____

_____

_____

(iii) Determine the resultant field strength at **X**. Use the space to the right of the diagram for a vector
      diagram, if required.   [3]

_____

_____

_____

_____

(b) (i)   Calculate the potential at point **X**.   [3]

_____

_____

_____

_____

(ii) A proton is released from point **X**. Calculate the maximum *speed* that it attains, giving your
      reasoning. [Proton mass = 1.67 × 10^{-27} kg]   [3]

_____

_____

_____

_____

Q6 Three equipotential surfaces are shown for an isolated positive point charge, Q.

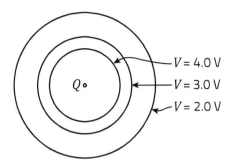

Explain why:

(a) The surfaces are spherical. [1]

....

....

(b) The surfaces are not equally spaced apart, even though their potentials differ by the same size step (1.0 V). [3]

....

....

....

....

....

Q7 (a) Explain why gravitational potentials are given as negative quantities. [The explanation is not the negative sign in $V = -GM/r$.] [2]

....

....

....

(b) A rocket is launched vertically from Mars at a velocity of 3000 m s^{-1}.

(i) Explain why it is necessary to use the equation, PE = $-GMm/r$, rather than the equation Δ(PE) = mgh, in relation to the rocket's subsequent motion. [2]

....

....

....

(ii) Calculate the maximum height above the Martian surface the rocket will reach. [4]
[Mass of Mars = 6.42×10^{23} kg; Diameter of Mars = 6780 km]

....

....

....

....

Q8 Point charges, $+Q$ and $-Q$, are placed a distance $2d$ apart, as shown:

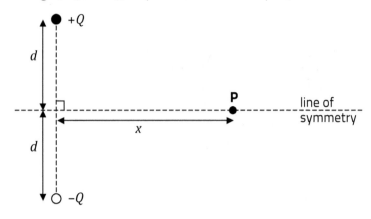

(a) (i) Determine the *electric field strength* at point P, in terms of Q, ε_0, d and x. You may add to the diagram. [4]

_____

_____

_____

_____

_____

_____

_____

(ii) Adam says that the potential at P is zero. Bethan says that this can't be right because work has to be done bringing a test charge along the line of symmetry from infinity to P. Evaluate who is right, explaining why the other is wrong. [3]

_____

_____

_____

_____

(b) **Compare** the variation in the electric field strength for the +/- charges with distance (from zero up to a large distance) **with** the case of two equal positive charges arranged in the same way. [4]

_____

_____

_____

_____

_____

_____

_____

Question and mock answer analysis

Q&A 1

(a) State what is meant by the *gravitational potential* at a point. [2]

(b) Two stars, A and B (see diagram), are much closer to each other than to any other stars.

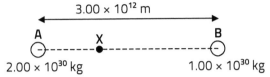

In the graph the gravitational potential, V, along a line joining A and B is plotted against r_A, the distance from star A. The distance **AX** is 1.00×10^{12} m.

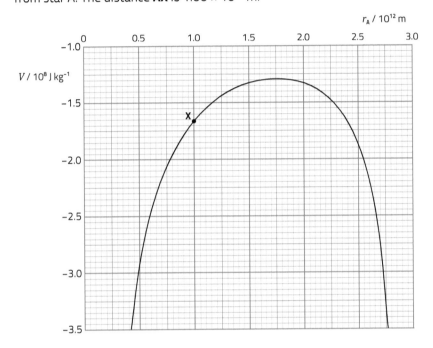

(i) Using data from the diagram, show clearly that point **X** on the graph is correct. [3]

(ii) **Use the graph** to determine the **resultant field strength** at X. [4]

(iii) Determine **from the graph** the value of r_A for which the resultant gravitational field strength is zero, giving a brief justification. [2]

(iv) Explain **in terms of field strength or force** why the point at which the resultant field strength is zero is closer to B than to A. Calculations calling for the use of a calculator are not needed. [2]

What is being asked

Part (a) serves two purposes: it tests that you know a standard AO1 definition and it sets the scene for part (b).

Part (b) consists mainly of AO2 as you are applying given data.

In (b)(i), rather than asking you to plot your own graph (taking several minutes of your time) the examiner finds out just as much about your understanding of the physics involved by asking you to check a single point. You must, though, present your check clearly.

(b)(ii) requires the use of a specific relationship and a specific technique. You must use the graph (as instructed in bold!) even if another method occurs to you.

In (b)(iii) it's a pretty good guess that one of the marks will be for the 'brief justification' leaving only one mark for the actual value, so it can't require much labour to determine!

(b)(iv) departs from *potential*, the main theme of the question. Although numerical calculations are not needed, you are allowed to use equations. Words will also be required!

Mark scheme

Question part		Description	AOs			Total	Skills	
			1	2	3		M	P
(a)		The work done [by an external force] taking a [test] mass from infinity to the point [1] divided by the [test] mass. [1]	2			2		
(b)	(i)	$V_A = -\dfrac{6.67 \times 10^{-11} \times 2.00 \times 10^{30}}{1.00 \times 10^{12}}$ [1] = -1.334×10^8 J kg^{-1} $V_B = -\dfrac{6.67 \times 10^{-11} \times 1.00 \times 10^{30}}{2.00 \times 10^{12}}$ [1] = -0.334×10^8 J kg^{-1} $V_A + V_B = -1.67 \times 10^8$ J kg^{-1} ecf **and** comment that this agrees with graph [or, to obtain ecf credit, doesn't agree]. [1]		3		3	1	
	(ii)	Tangent drawn to curve at **X** [1] In gradient equation 'rise' and 'run' put in correctly, but tolerating slips in powers of 10. [1] g from 1.1×10^{-4} to 1.3×10^{-4} N kg^{-1} **unit** [1] g stated to be towards **A**. Accept to the left. [1]		4		4	4	
	(iii)	$r_A = 1.75 \times 10^{12}$ [m] [$\pm 0.05 \times 10^{12}$ m] [1] Gradient is zero here, or turns from + to –. [1]	1	1		2	1	
	(iv)	For zero resultant field: $\dfrac{[G]M_A}{r_A^2} = \dfrac{[G]M_B}{r_B^2}$ **or** inverse sq. law referred to [1] $M_B < M_A$ so $r_B < r_A$ **or** clear argument in words [1]		2		2		
Total			3	10	0	13	6	

Rhodri's answers

(a) The work done when a body is taken from infinity to the point.✓ X

MARKER NOTE
Rhodri has the main thrust of the definition correct, but the omission of 'per unit mass' is a serious mistake. He has, in effect, defined potential energy rather than potential.

1 mark

(b) (i) V due to $A = -\dfrac{6.67 \times 10^{-11} \times 2.00 \times 10^{30}}{1.00 \times 10^{12}}$ ✓

$= -1.33 \times 10^8$

V due to $B = -\dfrac{6.67 \times 10^{-11} \times 1.00 \times 10^{30}}{2.00 \times 10^{12}}$ ✓

$= -0.33 \times 10^8$

$V_A - V_B = -1.00 \times 10^8$ X

The agreement with the graph (-1.7×10^8) is not very good X no bod

MARKER NOTE
Rhodri has calculated V_A and V_B correctly, but has *subtracted* V_B from V_A, probably confusing the scalar V with the vector g.

2 marks

(ii)

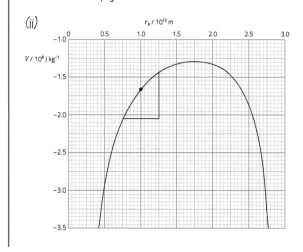

X X [No tangent drawn]

$$g = \frac{-1.45 \times 10^{8} - (-2.05 \times 10^{8})}{1.25 \times 10^{12} - 0.75 \times 10^{12}}$$

$$= 1.2 \times 10^{-4} \text{ N kg}^{-1} \checkmark \text{ bod}$$

Field is towards star A. $\checkmark$

MARKER NOTE

Rhodri has not drawn a tangent and loses the first two marks. The construction he has used produces an approximation to the gradient at 1.0×10^{12} m. He has actually calculated the mean value of g between 0.75 and 1.25×10^{12} m. In fact his value of g lies within the permitted range, and he has remembered to give the direction of this vector, so gains the last two marks.

2 marks

(iii) The resultant field is zero when r_A is 1.75×10^{12} m $\checkmark$ because the graph is highest here. $\times$

MARKER NOTE

Rhodri's value of r_A is correct, but his attempted justification doesn't connect with graph gradient.

1 mark

(iv) Because star B is lighter, we need to be closer to it to get a field strength from it that is equal and opposite to the field strength from A. $\times$ $\checkmark$

MARKER NOTE

Rhodri has a good feel for what's going on, and certainly deserves the 2nd mark, but he hasn't explained *why* being closer to B will compensate for B's smaller mass – hence he loses the 1st mark.

1 mark

Total   **7 marks /13**

Ffion's answers

(a) The work done per unit mass by the gravitational force when a mass goes from the point to infinity. $\checkmark\checkmark$

MARKER NOTE

Ffion's answer is equivalent to the standard definition given in the mark scheme.   **2 marks**

(b) (i) $V_A + V_B =$
$$-6.67 \times 10^{-11} \times \left(\frac{2.00 \times 10^{30}}{1.00 \times 10^{12}} + \frac{1.00 \times 10^{30}}{2.00 \times 10^{12}} \right)$$
$$= 1.67 \times 10^{8} \text{ J kg}^{-1} \checkmark\checkmark$$

This is exactly the value of V from the graph. $\checkmark$

MARKER NOTE

Ffion's working is clear and economical. She has clearly done this sort of calculation before!   **3 marks**

(ii)

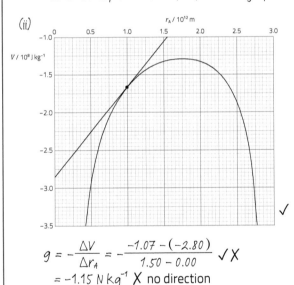

$$g = -\frac{\Delta V}{\Delta r_A} = -\frac{-1.07 - (-2.80)}{1.50 - 0.00} \checkmark \times$$
$$= -1.15 \text{ N kg}^{-1} \times \text{ no direction}$$

MARKER NOTE

Ffion has drawn a good tangent. 1st mark gained. She has found the gradient correctly, apart from missing powers of 10. 2nd mark gained, 3rd lost. She has remembered that g is *minus* the potential gradient, but has not interpreted the minus sign in her answer in terms of a *direction*. 4th mark lost.   **2 marks**

(iii) $r_A = 1.75 \times 10^{12}$ m $\pm 0.05 \times 10^{12}$ m $\checkmark$, as the gradient is zero somewhere in this region. $\checkmark$

MARKER NOTE

This is a very good answer. It was a nice touch to give an uncertainty, though there is no credit for it in the mark scheme.   **2 marks**

(iv) For the field strengths from the two stars to cancel to zero, their magnitudes must be equal, but $g \propto m/r^2$, so for the same g, a smaller m requires a smaller r, that is the point of zero resultant is closer to B than A. $\checkmark\checkmark$

MARKER NOTE

Ffion's explanation is clearly and logically argued.   **2 marks**

Total   **11 marks /13**

Section 3: Orbits and the wider universe

Topic summary

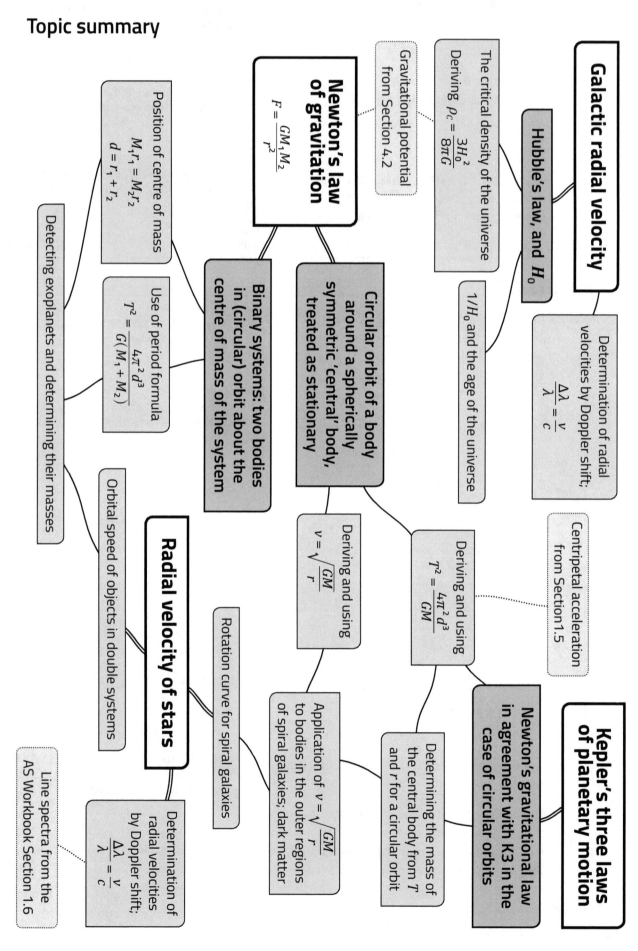

Q1 The Earth's orbit of the Sun is almost circular, with a radius of 150×10^6 km. Calculate a value for $M_\odot$, the mass of the Sun. [3]

....

....

....

....

....

Q2 (a) Assuming that the Earth is a sphere of radius 6370 km, show clearly that its mass is approximately 6 $\times 10^{24}$ kg. Use standard value for G and for g at the Earth's surface. [3]

....

....

....

....

....

(b) For your calculation in (a) to be valid, what do we have to assume about the way the Earth's mass is distributed? [1]

....

(c) Using your answer to (a), calculate a value for the mean density, ρ_{Earth}, of the Earth. [2]

....

....

....

....

Q3 A student is given a sketch of a comet's orbit around a star, and is asked to mark the positions of the star (S) and the point (X) in its orbit at which the planet moves most slowly. Here is his attempt:

Discuss whether the student's positions for S and X could be correct. [3]

....

....

....

....

....

Q4 (a) Astronomers commonly use the *astronomical unit* (AU) as a unit of distance in the solar system. This is the mean distance from the Earth to the Sun. Show that 1 AU is approximately 150 million km. [Mass of Sun = 1.99×10^{30} kg.] [2]

(b) Show that the light year (the distance travelled by light in one earth year) is approximately 9.5×10^{12} km. [2]

Q5 (a) The Moon orbits the Earth in an almost circular orbit of radius 3.83×10^5 km. Its period of revolution is 27.3 days. Hence show that the pull of the Earth must provide it with a centripetal acceleration of approximately 3×10^{-3} m s^{-2}. [3]

(b) (i) Hence calculate the ratio:
$$\frac{\text{acceleration due to Earth's gravity at Earth's surface}}{\text{acceleration due to Earth's gravity at Moon's distance}}$$
[1]

(ii) Treating the Earth as a sphere of radius 6.37×10^6 m, calculate the ratio:
$$\left(\frac{\text{distance of Moon from Earth's centre}}{\text{distance of Earth's surface from Earth's centre}}\right)^2$$
[1]

(c) Explain how these results support Newton's law of gravitation. [Newton did the equivalent calculations with the data available in his day.] [2]

Q6 (a) Calculate the height above the Earth's surface at which a geostationary satellite must orbit.
[Earth's radius = 6.37×10^6 m, Earth's mass = 5.97×10^{24} kg.] [4]

(b) State another requirement of the orbit for the satellite to be geostationary. [1]

Q7 Here are data for the (nearly circular) orbits of the two moons of Mars.

	Radius / 10^6 m	Period / day
Deimos	23.46	1.263
Phobos	9.39	0.319

Evaluate whether or not Kepler's 3rd law applies to these moons. [3]

Q8 An equation for the critical density of a 'flat' universe is:

$$\rho_c = \frac{3H_0^2}{8\pi G}$$

(a) Explain what is meant by *critical density*. [1]

(b) Show that the equation is correct as far as dimensions (or SI units) are concerned. [1]

Q9 The wavelength of an absorption line in a star's spectrum is found to vary regularly between 393.14 nm and 393.82 nm. Its wavelength as measured in the laboratory is 393.36 nm.

(a) Calculate the extreme values of the radial velocity of the star. [3]

....

....

....

....

....

(b) Without further calculation, describe a likely way in which the star is moving. Give reasons for your answer. [3]

....

....

....

....

....

Q10 Two stars, S_1, of mass 1.5×10^{30} kg, and S_2, of mass 2.5×10^{30} kg, travel in circular orbits about their common centre of mass, C. The stars are separated by 3.0×10^{12} m.

Calculate:

(a) The periodic time. [3]

....

....

....

....

(b) The radii of the orbits. [2]

....

....

....

....

Q11   Radial velocity is plotted against time for the star Tau Boötis A.

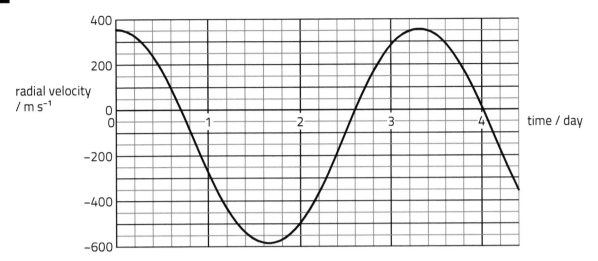

Assuming that its orbit is circular and seen edge-on:

(a) Show that the orbital velocity is approximately 500 m s^{-1}.   [2]

(b) Calculate the radius of the orbit.   [2]

(c) The mass of Tau Boötis A is estimated to be 2.6 × 10^{30} kg. Its orbital motion is due to an unseen planet of much smaller mass. Calculate an approximate value for the radius of the planet's orbit.   [3]

(d) Calculate a value for the planet's mass.   [2]

Q12 The mass, m_{vis}, of a visible star is estimated as 12×10^{30} kg. The variation from the mean of its radial velocity is plotted against time.

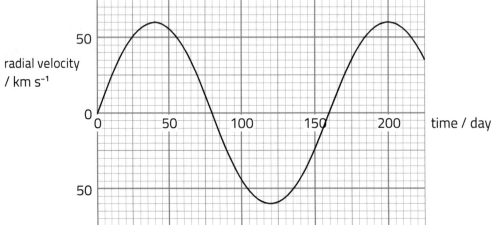

(a) Assuming that the orbit is circular and viewed edge-on, calculate its radius, r_{vis}. [2]

....

....

....

(b) The visible star is in mutual orbit with an invisible companion, presumed to be a black hole (BH). Explain why:

$$\left(\frac{2\pi}{T}\right)^2 r_{BH} = \frac{Gm_{vis}}{(r_{vis} + r_{BH})^2}$$

in which r_{BH} is the orbital radius for the black hole. [2]

....

....

....

....

(c) The only unknown in the equation in (b) is r_{BH}. But the equation is hard to solve. Check that the equation is satisfied approximately by $r_{BH} = 8.4 \times 10^{10}$ m. [2]

....

....

....

....

(d) Calculate the mass of the black hole. [2]

....

....

....

....

Q13 The diagram plots the radial velocities of galaxies against their distance away.
The distances are measured in megaparsec (Mpc). 1 Mpc = 3.09×10^{22} m.

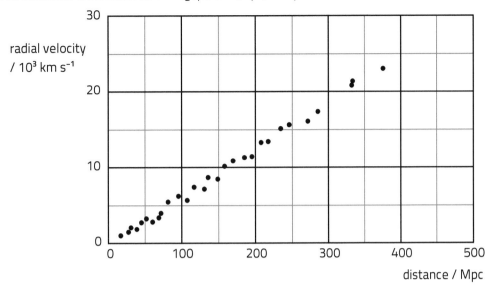

(a) Calculate a value for the Hubble constant, H_0, in SI units. [4]

....

....

....

....

....

....

(b) Give one reason, not involving errors of measurement, for the scatter of points. [1]

....

....

Q14 In a binary system, the two stars are separated by 30 AU and orbit the system's centre of mass in circular orbits of period 82.2 year. The orbital radius of the more massive star is 7.5 AU. Calculate the masses of the individual stars in terms of the solar mass, $M_\odot$.

[1 AU = mean distance from the Earth to the Sun = 1.50×10^{11} m; $M_\odot = 1.99 \times 10^{30}$ kg] [5]

....

....

....

....

....

....

....

....

....

Question and mock answer analysis

Q&A 1

(a) A body of mass, m, is in circular orbit of radius r around a spherically symmetric body of mass, M. [$M >> m$.] Starting from Newton's law of gravitation, show clearly that the orbital speed, v, of the body is given by: $v = \sqrt{\dfrac{GM}{r}}$ [2]

(b) The great majority of the stars in a galaxy, G, are contained in its 'central region', whose mass can be estimated from its total luminosity. This estimated mass has been used with the equation of part (a) to plot the **broken** curve on the graph. This shows how the speed of orbiting matter in the outer region of the galaxy is expected to depend on its distance, r, from the centre of the galaxy.

The **full** curve shows how the *observed* speed of orbiting matter depends on r.

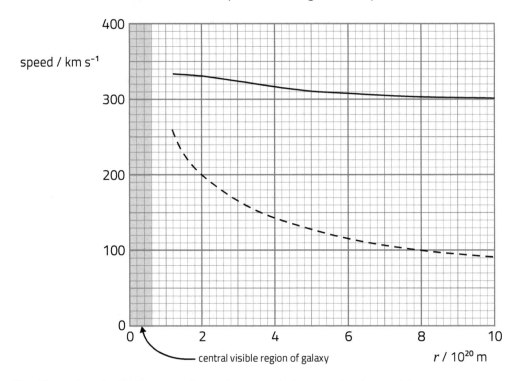

(i) Show that the **broken** curve is consistent with the equation in part (a). [2]

(ii) Determine the estimated mass of the galaxy, on which the **broken** curve is based. [3]

(iii) Explain what the two graphs suggest about the *actual* mass of the galaxy and the way in which the mass is distributed. [4]

(iv) Briefly describe one way that has been suggested to account for these apparent discrepancies regarding galactic mass. [1]

(c) Explain how the equation:

$$\frac{v}{c} = \frac{\Delta\lambda}{\lambda}$$

can be used to determine the speed of a body in a distant galaxy. [3]

What is being asked

(a) Here we have a quick and straightforward application of Newton's law of gravitation, in this piece of AO1 bookwork. 'Show clearly' means give some understandable working.

(b) (i) An AO3 question, with no hint as to how to proceed.

   (ii) This tests that you can put data from the graph into the equation of part (a), rearranging as necessary. Two or three rather basic skills are needed, but it's easy to make mistakes in reading from a graph!

   (iii) and (iv) These are AO1 questions. The implication of the observed variation of orbital speed of objects in the outer parts of a galaxy is expected in the specification. The difficulty here is in producing concise answers including an explanation of how they link to the data.

(c) Clearly you must state what the symbols in the equation mean, and which values are found by measurement. Do give details – but be aware of your limited time.

Mark scheme

Question part		Description	AOs			Total	Skills	
			1	2	3		M	P
(a)		$\dfrac{mv^2}{r} = \dfrac{GMm}{r^2}$ **or** $mr\omega^2 = \dfrac{GMm}{r^2}$ [1] Any correct intermediate algebraic step. [1]	2			2	1	
(b)	(i)	Two data points chosen, e.g. (2, 200), (8, 100) [1] Convincing manipulation $\longrightarrow v \propto r^{-0.5}$ [1]			2	2	1	
	(ii)	$M = \dfrac{rv^2}{G}$ (rearrangement at any stage) [1] Recognisable readings with correct powers of 10 taken from a point on the broken curve. [1] $M = 1.2$ or 1.3×10^{41} kg [1] Accept 3 sf: No *second* penalty for wrong powers of 10		3		3	1	
	(iii)	Actual mass greater than estimated [1] because speed higher [throughout] [1] Mass extends out [further than supposed] from centre. [1] Because speed hardly decreases with distance from centre, or equiv. [1]	4			4		
	(iv)	**Either** Dark matter with some explanation, e.g. not emitting light (accept 'hidden') or throughout galaxy **or** Newton's law of gravitation fails. [1]	1			1		
(c)		Measure wavelength of light from the body [1] In the equation, $\Delta\lambda$ = shift in wavelength, λ = [expected] wavelength, v = speed of body. [1] One extra detail, e.g. [identifiable] emission **or** absorption line used, **or** galaxy needs to be edge-on, **or** positive $\Delta\lambda$ shows body moving away from us. [1]	3			3		
Total			10	3	2	15	3	

Rhodri's answers

(a) $\dfrac{GMm}{r^2} = mr\omega^2$ ✓

$\dfrac{GM}{r^2} = r\dfrac{v^2}{r}$ ✗

$v^2 = \dfrac{GM}{r^2}$

I don't know why this is wrong.

> **MARKER NOTE**
> Rhodri has chosen the formula for centripetal acceleration that is less convenient for the problem in hand. But his first equation is correct and scores the 1st mark. No more marks, though, because, in trying to get rid of ω, he confused $\omega = v/r$ with $a = v^2/r$.
> **1 mark**

(b) (i) When r doubles from $2 \rightarrow 4 \times 10^{20}$ m the speed drops from 200 to 140 km s^{-1} ✓ This is approximately half so there is inverse proportion. ✗

> **MARKER NOTE**
> Rhodri has hit the 1st marking point, in comparing two data points. He has not understood the significance of the square root for this answer and so misses the 2nd mark.
> **1 mark**

(ii) When $r = 2 \times 10^{20}$, $v = 200$ ✓

So $200^2 = \dfrac{6.67 \times 10^{-11} M}{2 \times 10^{20}}$ ✓

$M = \dfrac{200^2 \times 2 \times 10^{20}}{6.67 \times 10^{-11}} = 1.2 \times 10^{35}$ kg ✗

> **MARKER NOTE**
> Rhodri's answer is correct except for not taking account of the km s^{-1} unit on the vertical scale. This is a common type of error, but the rather generous mark scheme penalises it only once. It's slightly long-winded to make M the subject *after* putting in the numbers, but whatever you're comfortable with...
> **2 marks**

(iii) The graph suggests that the mass of the galaxy is greater than that estimated. ✓ This follows from the equation in part (a) because the actual speed is much faster than that using the estimated mass, ✓ and the shapes of the graphs are different. [not enough]

> **MARKER NOTE**
> Rhodri has understood that there is more than the estimated mass and has justified it well, by appealing to the graph and the equation. He has not developed his remark about the shapes of the graphs to say anything useful about the mass distribution.
> **2 marks**

(iv) One idea is that there is hidden mass called 'dark matter'. 'Hidden' means that it doesn't interact with e-m radiation, which is why we wouldn't be able to detect it. ✓ Dark matter may consist of little-understood particles such as WIMPs.

Another suggestion is that there isn't really any extra mass, but that Newton's law of gravitation doesn't always work.

> **MARKER NOTE**
> Rhodri is clearly well-informed about dark matter! His answer is almost too thorough for the 1 mark allocated, and there was certainly no need for two different suggested solutions. In this case, as fairly generally, there is no penalty for giving too much information.
> **1 mark**

(c) $\Delta\lambda$ is the Doppler shift ✓ [bod] of the light and λ is its wavelength. c is the speed of light. The equation gives the velocity v of the body, which is away from us if the wavelength is increased. ✓

> **MARKER NOTE**
> Rhodri lost the 1st mark because he didn't actually say where the light comes from! The 2nd mark is unsafe because 'Doppler shift' doesn't specify *wavelength* shift, but the examiner gave benefit of doubt because of the reference to increase in *wavelength* at the end. The last mark is gained for the 'extra detail' about the direction of the velocity.
> **2 marks**

| Total | 9 marks /15 |

Ffion's answers

(a) $\dfrac{mv^2}{r} = \dfrac{GMm}{r^2}$ ✓

$v^2 = \dfrac{GM}{r}$ (times by $\dfrac{r}{m}$) ✓

$v = \sqrt{\dfrac{GM}{r}}$

MARKER NOTE
Ffion's answer is clear and correct. She has even given a commentary on her algebra!

2 marks

(b) (i) If $v = \sqrt{\dfrac{GM}{r}}$, then $v^2 \propto \dfrac{1}{r}$ so $\left(\dfrac{v_1}{v_2}\right)^2 = \dfrac{r_2}{r_1}$.

Take $r_1 = 2$; $v_1 = 200$ and $r_2 = 8$; $v_2 = 100$ ✓
$(r_2/r_1) = 4$ and $(v_1/v_2)^2 = 2^2 = 4$. These are the same so OK. ✓

MARKER NOTE
Ffion homes in on two easily used data points, with one radius 4× the other for the 1st mark. Her clear algebra identifies a test and she uses it correctly for the 2nd mark.

2 marks

(ii) Rearranging the given equation

$M = \dfrac{rv^2}{G}$ ✓

Putting in values

$M = \dfrac{4 \times 10^{20} \times 180\,000^2}{6.67 \times 10^{-11}}$ ✓ $= 1.94 \times 10^{41}$ kg ✗

MARKER NOTE
Ffion's method is crystal clear, but she misread the vertical scale, leading to a seriously inaccurate answer. There was only one mark deducted, but a meaner mark scheme could have resulted in a 2-mark loss.

2 marks

(iii) At all values of r in the outer galaxy, the actual speed is greater than the calculated speed, ✓ so if the equation in part (a) is true, the actual mass must be greater than the estimated mass. ✓ Also, the speed hardly falls with increasing distance, as it would if almost all the mass were in the central region (v proportional to $1/\sqrt{r}$). ✓ So it looks as if there is much more mass than assumed in the outer regions. ✓

MARKER NOTE
Ffion has produced what amounts to a model answer. Both the larger than estimated mass and the wider distribution of mass are noted and justified from the graphs, and equation $v \propto \dfrac{1}{\sqrt{r}}$ was a nice touch!

4 marks

(iv) The galaxy contains material (particles) which we cannot detect with electromagnetic radiation, but which has mass. It is called `dark matter'. ✓

MARKER NOTE
This is a clear and concise answer.

1 mark

(c) We examine light coming to us from the body. If we can recognise a wavelength as coming from a particular sort of atom ✓, and that wavelength is shifted by $\Delta\lambda$ from its usual value, λ, ✓ then the body is moving away from us with radial velocity $c \times \Delta\lambda/\lambda$, if $\Delta\lambda$ is positive. ✓

MARKER NOTE
Ffion's summary is very neat. She has explained the use of the equation and her reference to light coming from a particular sort of atom is a valuable extra detail.

3 marks

Total **14 marks /15**

Section 4: Magnetic fields

Topic summary

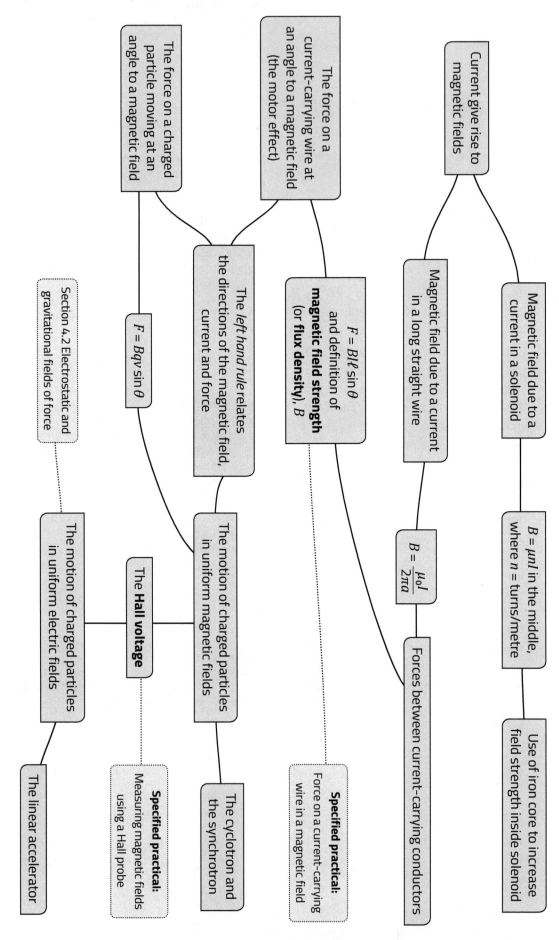

Current give rise to magnetic fields

The force on a current-carrying wire at an angle to a magnetic field (the motor effect)

The force on a charged particle moving at an angle to a magnetic field

Magnetic field due to a current in a long straight wire

Magnetic field due to a current in a solenoid

$F = BI\ell \sin\theta$ and definition of **magnetic field strength (or flux density)**, B

$F = Bqv \sin\theta$

The *left hand rule* relates the directions of the magnetic field, current and force

$B = \dfrac{\mu_0 I}{2\pi a}$

$B = \mu n I$ in the middle, where n = turns/metre

Section 4.2 Electrostatic and gravitational fields of force

Specified practical: Force on a current-carrying wire in a magnetic field

Forces between current-carrying conductors

Use of iron core to increase field strength inside solenoid

The motion of charged particles in uniform electric fields

The **Hall voltage**

The motion of charged particles in uniform magnetic fields

Specified practical: Measuring magnetic fields using a Hall probe

The cyclotron and the synchrotron

The linear accelerator

Q1   When placed in a magnetic field, B, a current-carrying wire experiences a force, F, given by:

$$F = BI\ell \sin\theta$$

(a)   Identify the quantities I, ℓ and θ.   [1]

....

....

(b)   Use the equation to express the tesla (T) in terms of the base SI units kg, m, s and A.   [2]

....

....

....

(c)   State the name of the rule that links the direction of F with those of the current and field.   [1]

....

Q2   A clockwise current of 2.5 A is maintained in a triangular loop of wire by a battery (not shown). A uniform magnetic field of 0.030 T, at right angles to the plane of the loop, is applied as shown:

(a)   Add arrows to the centres of each side of the loop to show the directions of the forces on the sides due to the uniform magnetic field.   [2]

(b)   Calculate the magnitudes of the forces on the following sides: [5]

(i)   AB ....

....

(ii)   BC ....

....

(iii)   CA ....

....

(c)   Show that the resultant force on the loop is zero.   [3]

....

....

....

....

....

Unit 4 Practice questions

Q3 A uniform magnetic field of magnitude B is applied to the right as shown, so that it surrounds the triangular loop shown, which carries current I.

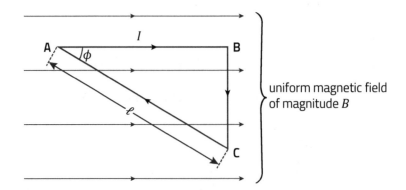

(a) Ella suggests that the resultant force on the loop is zero. Evaluate her suggestion. [4]

....

....

....

....

....

....

....

(b) Even so, Ella argues, the loop will not remain stationary unless it is fixed. Discuss this further claim. [2]

....

....

....

Q4 Three long, parallel wires, P, Q and R are 60 mm apart from each other, as shown in the (cross-sectional) diagram. They each carry a current of 7.5 A into the page.

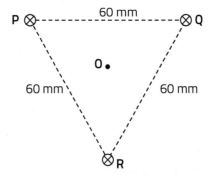

(a) (i) Point O is equidistant from P, Q and R. **Draw arrows labelled P, Q and R on the diagram** to show the directions of the magnetic fields at O due to P, to Q and to R. [2]

(ii) State the magnitude of the resultant magnetic field at point O, justifying your answer briefly.   [2]

....

....

....

(b) Stating its direction, determine the force on a 2.0 m length of wire R:

(i) when the current in wire Q is temporarily turned off;   [2]

....

....

....

(ii) when the current is restored in Q, so P, Q and R all carry 7.5 A again.   [3]

....

....

....

Q5 The diagram shows part of the magnetic field due to a current-carrying solenoid.

(a) The direction of the field in the solenoid is from left to right.

(i) List what else can be deduced about the solenoid's field from the pattern of magnetic field lines.   [3]

....

....

....

....

(ii) Describe how you could investigate the deductions you have made in part (i).   [3]

....

....

....

....

....

(b) State the direction, A or B, of the current, naming the rule that you have used.   [1]

....

(c) The solenoid is 60.0 cm long and has 300 turns of wire. Calculate the magnetic flux density, B, at its centre when it carries a current of 4.0 A. [2]

....

....

....

(d) State how you could increase the flux density at the centre, without changing the current, the number of turns or the length of the solenoid. [1]

....

Q6 The diagram shows a beam of electrons passing through a region of uniform magnetic field. The electrons are moving at a speed of 3.0×10^7 m s^{-1} in an arc of a circle of radius 0.040 m.

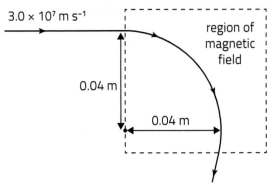

(a) The electrons have been accelerated to this speed from rest. Calculate the accelerating voltage used. [2]

....

....

....

(b) Determine the magnitude and direction of the magnetic flux density. [3]

....

....

....

(c) The electron beam can be made to go straight by applying, as well as the magnetic field, a suitable electric field. Determine its magnitude and direction. [2]

....

....

....

Q7 (a) A particle of mass m and charge q moves in a circular path in a uniform magnetic field, B, perpendicular to the plane of the circle. Starting from the force experienced by the particle, show that the time the particle takes per revolution is:

$$T = \frac{2\pi m}{qB}$$

[3]

....

....

....

....

....

(b) A simplified diagram of a cyclotron is given below:

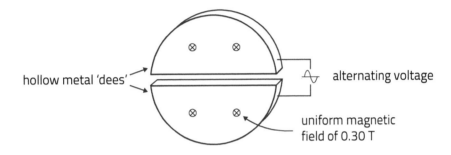

Calculate the frequency of alternating voltage needed if protons are being accelerated in the cyclotron and a magnetic field of 0.30 T is being applied. [2]

....

....

....

(c) A synchrotron is a particle accelerator whose principle was developed from that of a cyclotron. State two differences between a synchrotron and a cyclotron, and one difference between the ways in which they are used. [3]

....

....

....

....

....

....

Unit 4 Practice questions

Q8 Here is a simplified diagram of part of a linear accelerator for accelerating protons:

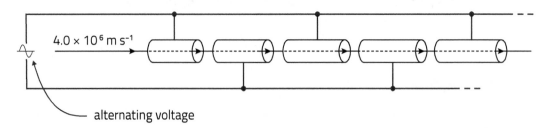

alternating voltage

Protons enter the left-hand tube at a speed of 4.0×10^6 m s^{-1}. As a proton passes from left to right across the gap between two tubes, the accelerating voltage between the tubes is (−)120 kV.

(a) Calculate a proton's speed after it has entered the right-hand (fifth) tube shown in the diagram.
[Mass of proton = 1.67×10^{-27} kg] [4]

....

....

....

....

....

....

....

(b) Explain:

(i) Why an *alternating* voltage is needed. [2]

....

....

....

(ii) Why the tubes are of different lengths. [2]

....

....

....

(c) State one disadvantage of a linear accelerator over a cyclotron. [1]

....

Question and mock answer analysis

Q&A 1

(a) In a vacuum, moving charged particles travel in circles at constant speed when a uniform magnetic field is applied at right angles to the particles' initial velocity. In a slice of conducting material carrying a current, the charged particles follow straight paths shortly after a uniform magnetic field is applied, as in the diagram.

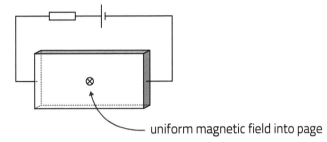

uniform magnetic field into page

Account for these different behaviours. Equations are not wanted. [6 QER]

(b) A tesla-meter consists of a probe containing a 'wafer' across which a Hall voltage is produced, and a 'meter unit' that supplies the wafer with a constant current, and also displays the measured value of the magnetic field.

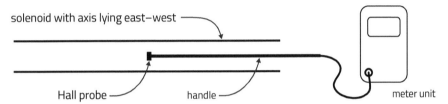

solenoid with axis lying east–west

Hall probe   handle   meter unit

Sally checks the calibration of the meter by placing its probe in the middle of a long solenoid with the probe orientated as shown, to produce the maximum reading.

The solenoid is 0.75 m long and has 150 turns of wire. Here are Sally's results:

solenoid current / A	0.0	0.2	0.4	0.6	0.8
meter reading / µT	3	53	104	154	204

(i) Suggest why Sally placed the solenoid with its axis lying East-West. [1]

(ii) Without drawing a graph, evaluate to what extent the results confirm that the tesla-meter is correctly calibrated. [3]

(c) Sally now uses the tesla-meter to investigate the magnetic field due to a long, straight, current-carrying wire. (See diagram.)

View from above

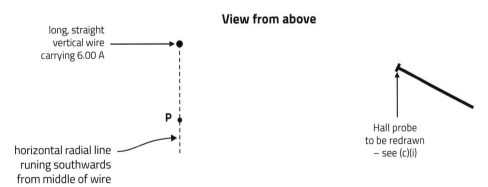

long, straight vertical wire carrying 6.00 A

P

horizontal radial line runing southwards from middle of wire

Hall probe to be redrawn – see (c)(i)

She places the probe at various points along the radial line, recording the meter reading and using a ruler to measure the distance, r, of the point from the centre of the wire.

(i) On the diagram redraw the Hall probe, positioned to measure the magnetic field at **P** due to the wire. [Note how the probe is positioned in part (a).] [1]

(ii) Sally's results are used to plot magnetic flux density, B, against $1/r$.

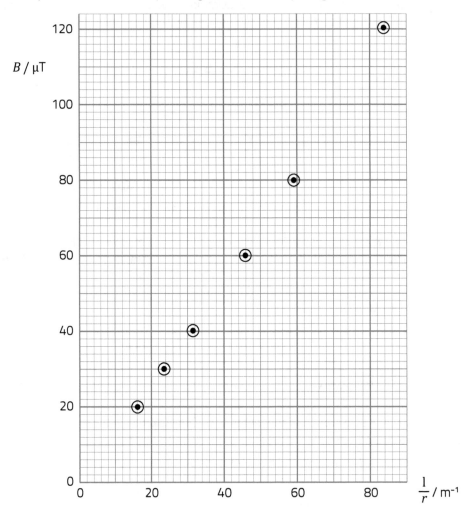

(I) Determine the current in the vertical wire. [4]

(II) Sally thinks that her values of r may be in error by a fixed amount because she has simply measured to the outer casing of the Hall probe. Her friend suggests that for this reason it would have been better to plot r against $\frac{1}{B}$. Discuss this suggestion. [2]

What is being asked

Part (a) is an AO1 question which combines two pieces of familiar bookwork. Even the diagram should be familiar. Hence AO1. The question looks for step-by-step explanation in a logical sequence with correct use of technical terms. Part (b)(i) is a straightforward AO2 mark but the mark allocation for (b)(ii) and 'to what extent' suggest there is more than one point to be made in this AO3 question.

Part (c)(i) tests knowledge of the shape of the wire's field with (c)(ii)I combining the analysis of a straight line with the expected equation for the wire's field – hence AO3. The last part is a more searching test of candidates' understanding of straight-line graph theory.

Mark scheme

Question part			Description	AOs			Total	Skills	
				1	2	3		M	P
(a)			**In a vacuum** • Magnetic force at right angles to velocity. • No increase in speed because no work done on particle or no force component parallel to velocity. • Force changes direction of particle velocity [at a constant rate]. **In conducting material** • Magnetic force deflects particles to the top (or bottom) of the slice. • Use of Left Hand Motor rule. • Displaced charged particles set up E field. • Force due to E opposes that due to B. • Equilibrium established when force due to E is equal (and opposite) to force due to B. • No resultant force so particles go straight.	6			6		
(b)	(i)		So Earth's magnetic field doesn't affect measurement, or equivalent. [1]		1		1		1
	(ii)		B calculated from $B = \mu_o nI$ for at least one value of I, e.g. for 0.20 A, $B = 50.3$ mT [1] All readings correct **or** proportional to current [1] if zero error (or constant error) of 3 mT subtracted [1]			3	3	3	3
(c)	(i)		Probe at **P**; handle horizontal		1		1		
	(ii)	(I)	Straight line drawn close to points *except top one*, and through origin or within 1 small square to its right. [1] Gradient determined or equivalent. Accept poor line and/or slips in powers of 10 for this mark. [1] $B = \dfrac{\mu_0 I}{2\pi r}$ used, e.g. gradient $= \dfrac{\mu_0 I}{2\pi}$ [1] $I = 6.5$ [±0.1] A or 6.50 [±0.10] A **Unit** [1] [ecf on line drawn]			4	4	2	4
		(II)	$1/B$ against r would produce a straight line [despite error in r] but B against $1/r$ wouldn't. [1] Error in r appears as [−] intercept on r axis [or equiv.] [1]			2	2	2	2
Total				6	2	9	17	7	10

Rhodri's answers

(a) In a vacuum the force acts at right angles to the particle's speed, and makes it go in a circle. In the conducting material the same force makes the charged particles build up on the top of the slice and this stops the magnetic force from deflecting the particles any more, so they go straight.

MARKER NOTE
Apart from using 'speed' instead of the vector, 'velocity', Rhodri has correctly stated the direction of the force and realised that it's what makes the path a circle in a vacuum. He hasn't told us why the speed is constant. He seems to have some limited understanding of why the particles follow straight lines in the material. But he gives no reason for particles being deflected upwards, and he doesn't clearly say that their build-up is responsible for a second force opposing the magnetic force. A low-band answer.

2 marks

(b) (i) So that the Earth's magnetic field doesn't interfere with the results. ✓

MARKER NOTE
Correct answer.

1 mark

(ii) The reading is almost proportional to the current because 104 is almost twice 53 and so on. ✓ But the field should be exactly proportional to the current, so the meter calibration is good but not perfect.

MARKER NOTE
Middle mark gained for spotting the approximate proportionality, but analysis far from complete. Rhodri hasn't realised the need to compare a calculated value B for the solenoid with the results. Nor has he spotted the constant error.

1 mark

(c) (i) [Probe drawn at P with handle along broken radial line] ✗

MARKER NOTE
The probe is orientated to measure a radial B but the field is tangential.

0 marks

(ii) (I)

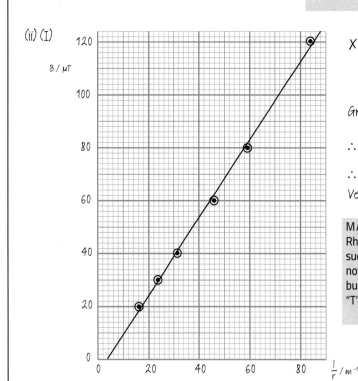

✗

Gradient $= \dfrac{98}{67} = 1.4626$ ✓

$\therefore 1.4626 = \dfrac{4\pi \times 10^{-7} \times I}{2\pi}$ ✓

$\therefore I = 7.313 \times 10^{6}$ A ✗

Very large current!

MARKER NOTE
Rhodri's line is quite close to all the points, but the three successive points below the line suggest that this was not the best choice. He then gains the middle two marks but loses the last because he missed the "μ" in front of "T" on the graph axis!

2 marks

(II) There will be no difference because $B = \dfrac{\mu_0 I}{2\pi r}$ and $r = \dfrac{\mu_0 I}{2\pi B}$ give lines of the same gradient. ✗

Except that an error in r will just shift the line up or down. ✓

MARKER NOTE
Rhodri's last sentence gains the second mark, but he hasn't made clear that, with zero errors in r, plotting B against $1/r$ doesn't give a straight line.

1 mark

Total **7 marks /17**

Ffion's answers

(a)   A charged particle in a magnetic field experiences a force at right angles to its velocity. This makes it change its direction continuously. It does this at a constant rate because the field is uniform. So it goes in a circle in a vacuum.

In the conducting material a Hall voltage builds up between top and bottom surfaces. This is because the particles are deflected upwards by the Left Hand Motor rule (if the particles are positive). The build-up of positive charge makes a downward force on the particles, opposing the force from the magnetic field. Soon the forces balance and the particles go straight (Newton's law 1).

MARKER NOTE
Ffion has accounted convincingly for the particle's circular motion in a vacuum, but not for the constancy of its speed.
She applied the Left Hand Motor rule correctly, and realised that the build-up of charge gives rise to a force that opposes the magnetic force. It would have been better still if she had mentioned the setting-up of an electric field, but her reference to the Hall effect and her correct use of Newton's first law confirm that this is a top band answer.

6 marks

(b)(i)   The probe will not pick up the Earth's field, which is northwards. ✓

MARKER NOTE
Correct answer – very clear.

1 mark

(ii)   For I = 0.20 A,
$B = 4\pi \times 10^{-7} \times (150/0.75) \times 0.20 = 50$ mT, ✓

and for the different currents used, B = 0, 50, 101, 151, 201, 251 mT. ✓

The readings are all 3 mT higher than these, so apart from a fixed error of 3 mT they are correct. ✓

MARKER NOTE
Although proportionality not mentioned, question answered fully.

3 marks

(c)(i)   [Probe drawn at P with handle at right angles to radial line] ✓

MARKER NOTE
Probe correctly orientated.

1 mark

(ii)(I)

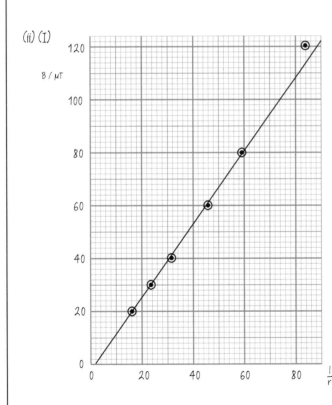

✓

$$B = \frac{4\pi \times 10^{-7} I}{2\pi} \times \frac{1}{r}$$

$$\therefore \text{Gradient} = \frac{4\pi \times 10^{-7} I}{2\pi} \checkmark$$

$$\therefore \frac{120 \times 10^{-6}}{84 - 2} = \frac{4\pi \times 10^{-7} I}{2\pi} \quad \times$$

$$\therefore I = 7.3 \text{ A} \checkmark \text{[ecf]}$$

MARKER NOTE
By ignoring the last point, Ffion's line fits the other points well. A good choice followed by competent analysis, though she has lost the last mark through reading 88 as 84 on the horizontal scale!

3 marks

(II)   If there is an error of a in r. Then using r for the measured value, $r - a = \frac{\mu_0 I}{2\pi B}$.

$\therefore r = \frac{\mu_0 I}{2\pi B} + a$. So now the line will be straight ✓, with a as the intercept, when r is plotted against 1/B. ✓

MARKER NOTE
Ffion has presented the theory of the proposed graph very clearly. Her use of the word 'now' implies that the original graph would not be straight.

2 marks

Total   16 marks /17

Section 5: Electromagnetic induction

Topic summary

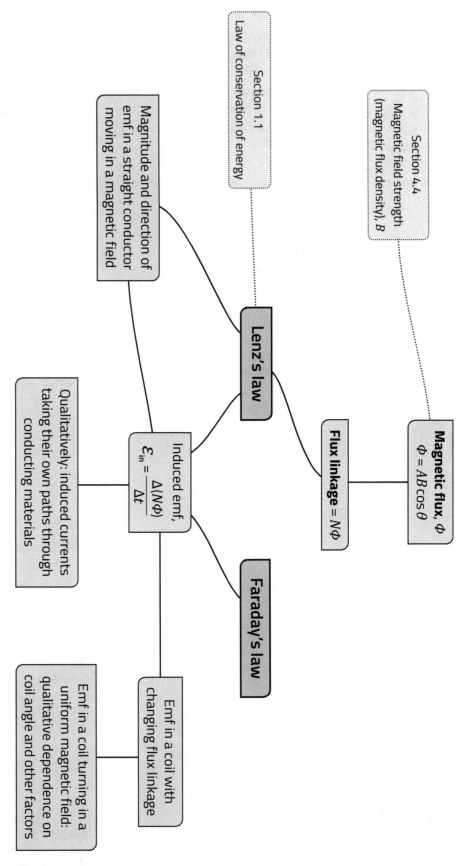

Q1 A ring made of copper wire has a **diameter** of 0.080 m. A uniform magnetic field of 0.050 T is applied to it, at right angles to the plane of the ring (into the paper in the diagram).

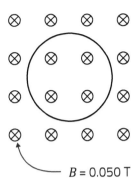

$B = 0.050$ T

(a) (i) Calculate the magnetic flux through the ring. [2]

....

....

....

(ii) Calculate the emf induced in the ring when the field strength is reduced to zero at a constant rate in a time of 0.16 s. [2]

....

....

....

(iii) State the sense of the induced emf (clockwise or anticlockwise in the diagram), justifying your answer clearly. [3]

....

....

....

....

....

(iv) The resistance of the ring is 2.75×10^{-3} Ω. Calculate the energy dissipated in the ring during the time that the field is changing. [2]

....

....

....

(b) Alice claims that doubling the diameter of the ring (but keeping the thickness of wire the same) would double the energy dissipated if the experiment of part (a) were repeated. [The magnetic field extends indefinitely.] Evaluate her claim. [2]

....

....

....

Unit 4: Practice questions

Q2 A square loop of wire, ABCD, measuring 0.15 m × 0.15 m is being pushed from left to right at a steady speed of 0.20 m s^{-1}.

(a) At the instant shown in diagram (a) the leading edge, **BC**, of the loop has entered a region of uniform magnetic field (bounded by the broken line). For this instant:

(i) Calculate the emf induced in the square loop. [2]

....

....

....

(ii) Calculate the current, given that the loop resistance is 0.020 Ω, **and show the direction of the current on diagram (a)**. [1]

....

....

(iii) Calculate the motor effect (magnetic) force that acts on **BC and show its direction on diagram (a)**. [3]

....

....

....

....

(b) Explain why there is no current in the loop as it passes through the position shown in diagram (b). [2]

....

....

....

(c) **Draw an arrow**, correctly positioned on diagram (c), to show the force acting on the loop when the leading edge, **BC**, has passed out of the field. [2]

Q3 (a) State Lenz's law of electromagnetic induction. [2]

....

....

....

(b) Using the set-up in question 2 (a) as your example, explain how Lenz's law is an application of Conservation of Energy. [2]

....

....

....

....

(c) Making use of your answers to question 2 (a), determine:

(i) The power dissipation by resistive heating in the loop. [2]

....

....

....

(ii) The work done per second pushing the loop into the field. [2]

....

....

....

....

Q4 A square coil of 150 turns measures 12.0 cm × 12.0 cm. It is placed so that the normal to the square is at an angle of 30° to the Earth's magnetic field at a place where its magnitude is 48 μT.

(a) Determine:

(i) The magnetic flux through the coil. [2]

....

....

....

(ii) The magnetic flux linkage. [1]

....

(b) The coil is now turned in a time of 1.2 s so that its normal is at right angles to the field. Calculate the mean emf induced in the coil. [2]

....

....

....

....

Q5 (a) State Faraday's law of electromagnetic induction. [2]

....

....

....

(b) The diagram is an edge-on view of a square coil, PQRS, which is being rotated at constant angular velocity, ω, about an axis passing through the midpoints of sides PQ and RS. The coil is in a uniform magnetic field, B, as shown. A resistor is connected across the coil.

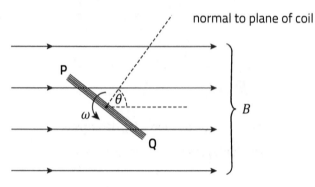

(i) Explain, using the concept of magnetic flux, how the magnitude of the **current** depends on angle θ, **and on other factors**. Equations are not wanted. [6 QER]

....

....

....

....

....

....

....

....

....

....

....

(ii) Sketch the variation of induced current with time over **two** complete revolutions of the coil. Start from the instant when $\theta = 0$. [2]

Q6

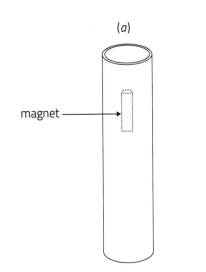

(a)   (b)   (c)

(a)   A magnet is dropped down a vertical copper tube (diagram (a)). It does not touch the walls of the tube. The magnet approaches a terminal speed much sooner than it would if falling in a non-conducting tube of the same dimensions.

Explain carefully how the extra force resisting the motion arises.   [3]

....

....

....

....

....

(b)   A saw is used to make a narrow cut in the wall of the tube all down its length (diagram (b)). Explain briefly why the magnet now falls much more freely down the tube.   [1]

....

....

(c)   The tube in (a) is now cut into rings, and these are glued back together with insulating glue (diagram (c)). Nabila believes that the magnet's fall through the re-assembled tube will now be more like that in (a) than that in (b). Discuss whether or not her belief is likely to be correct.   [2]

....

....

....

Q7 (a) A metal rod rests on metal rails a distance, l, apart, as shown. The circuit is completed by a resistor. There is a uniform magnetic field, B, directed into the paper.

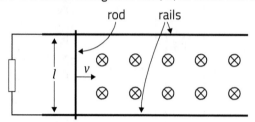

The rod is pushed to the right at a constant speed, v. A textbook gives two equations for the magnitude of the emf induced in this set-up:

$$E = Blv \quad \text{and} \quad E = \frac{\Delta\Phi}{\Delta t}$$

By considering the area swept out by the portion of rod between the rails in time Δt, show clearly that the two equations are equivalent. You may add to the diagram. [3]

....

....

....

....

(b) Starting at time $t = 0$, the rod shown in the left-hand diagram below is pushed to the right at a constant speed of 0.50 m s^{-1}. It makes contact with the conducting rails.

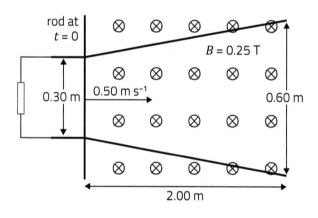

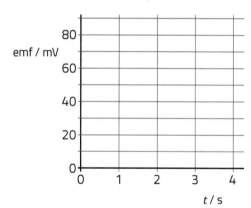

(i) Show that the emf initially is roughly 40 mV. [2]

....

....

(ii) **Use the graph grid** on the right (above) to show how the emf depends on time, t, over the interval $0 < t \leq 4$ s. [2]

(iii) The resistance of the resistor is 1.50 Ω. The resistances of the rod and rails are negligible. Determine the current at $t = 3.0$ s. [2]

....

....

....

Question and mock answer analysis

Q&A 1

(a)   A bar magnet is dropped through a flat circular coil, as shown in the side-view in Fig 1. The flux, Φ, through the coil varies with time, t, as shown in Fig 2.

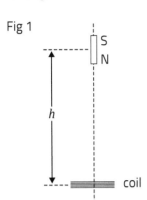

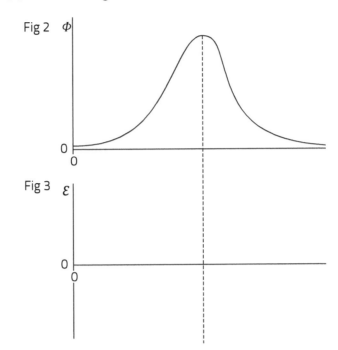

Fig 1

Fig 2

Fig 3

(i)   Explain briefly why the flux falls more quickly than it rises.   [1]

(ii)   On the axes in Fig 3, sketch a graph of the emf induced in the coil against time. Take the initial emf as positive.   [4]

(b)   Tom connects a flat circular coil to a meter adapted to record the maximum emf, $\mathcal{E}_{max}$, induced in the coil when the magnet is dropped through it. He believes that:

$$\mathcal{E}_{max} = kv$$

in which k is a constant and v is the magnet's speed at the instant of maximum emf.

He drops the magnet four times from a height, h, above the coil (see Fig 1), recording $\mathcal{E}_{max}$ each time. He repeats the procedure for five more values of h, and plots points for $\mathcal{E}_{max}$ against h, together with their error bars on the graph grid. [See next page].

(i)   Tom's values of $\mathcal{E}_{max}$ when $h = 0.400$ m are 6.0 mV, 6.1 mV, 6.3 mV, 6.1 mV.

Calculate the mean value and uncertainty in $\mathcal{E}_{max}^2$, and hence comment on whether the point and its error bar have been correctly plotted.   [4]

(ii)   Tom expects $\mathcal{E}_{max}$ and h to be related by the equation:

$$\mathcal{E}_{max}^2 = 2k^2 g\,(h + h_0)$$

where h_0 is a small constant distance.

Justify this equation.   [3]

(iii)   Use the graph to find a value for k, together with its absolute uncertainty.   [6]

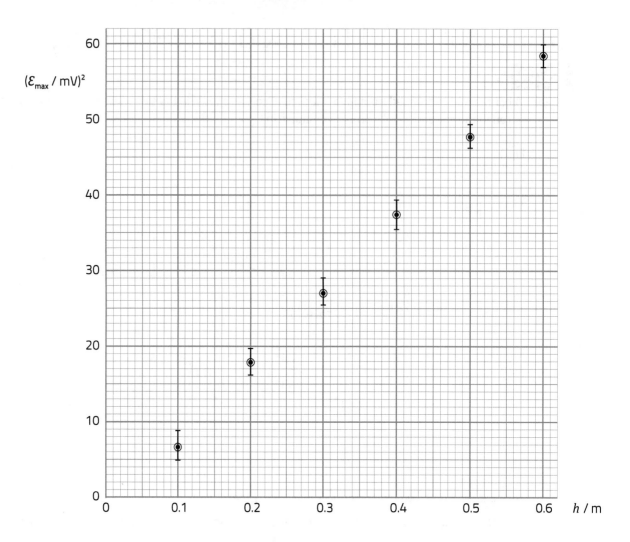

What is being asked

This is a fair question but sections (b)(ii) and (iii) are rather tough. This bit of the specification is conceptually demanding and this experiment is not one of the specified practicals. Having said that, (a)(i) is designed to be an easy AO2 mark, which draws the attention of the student to an important feature of the graph, whilst the mark allocation in (a)(ii) should alert the student to the fact that four features are expected. Part (b)(i) is a routine uncertainty question in which the presentation of results is important. Although the physics should be familiar in (b)(ii) the synoptic aspect adds to the difficulty. In (b)(iii) the procedure is not unique to the question but more steps are involved with no hints given, making this a rather testing AO3.

Mark scheme

Question part		Description	AO 1	AO 2	AO 3	Total	Skills M	Skills P
(a)	(i)	Magnet leaves coil more quickly than enters. [1]		1		1		
	(ii)	Double hump with a zero at time of peak F. [1] Second hump inverted. [1] Second hump deeper than first is tall. [1] Second hump more compressed in time. [1]		4		4		
(b)	(i)	Mean of $\mathcal{E}^2_{max}$ = 37.5 or 38[(mV)^2 or $\times 10^{-6}$ V^2] [1] % unc in $\mathcal{E}^2_{max}$ = 2.4 [or 2.5 or 2.45 or by impl] [1] Unc in $\mathcal{E}^2_{max}$ = 1.8 or 1.9 or 2 [(mV)^2 or $\times 10^{-6}$ V^2] [1] Final presentation with units and consistent decimal places and with sensible comment (ecf) on plotted point, e.g.... 37.5 ± 1.8 (mV)^2 or 38 ± 2 × 10^{-6} V^2 so plotted point and error bar correct. [1]		4		4	4	4
	(ii)	h_0 linked to maximum emf not being when centre of magnet level with coil. [1] For next 2 marks accept omission or mishandling of h_0. $v^2 = u^2 + 2ax$ or $\frac{1}{2}mv^2 = mg(h[+h_0])$ leading to.... ... so $v^2 = 2g(h[+h_0])$ [1] Substitution from $\mathcal{E}^2_{max} = kv$ so $\mathcal{E}^2_{max} = 2k^2(h[+h_0])$ [1]	1 (for $v^2 = 2g(h[+h_0])$); 1 (for $\mathcal{E}^2_{max} = 2k^2(h[+h_0])$)	1		3	1; 1	
	(iii)	At least one gradient evaluated using rise/run. Accept poor line and slips in power of 10 for this mark. [1] Max gradient between 107 × 10^{-6} [V^2 m^{-1}] and 111 × 10^{-6} [V^2 m^{-1}] [1] Min gradient between 95 × 10^{-6} [V^2 m^{-1}] and 99 × 10^{-6} [V^2 m^{-1}] [1] Penalise only once above for wrong power of 10. Correct use of gradient = $2 k^2 g$. [1] $k = 2.99 × 10^{-3}$ V m^{-1} s [or 3.0 × 10^{-3} V m^{-1} s] with ecf on gradients **UNIT** [1] ± 0.13 × 10^{-3} V m^{-1} s [or 0.1 or 0.2 × 10^{-3} V m^{-1}s] with ecf on gradients [1]			6	6	6	6
Total			2	10	6	18	12	10

Rhodri's answers

(a) (i) The magnet is falling faster as it leaves. ✓

(ii)

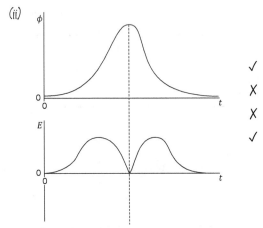

✓
X
X
✓

(b) (i) $\langle E_{max}^2 \rangle = \frac{1}{4}(36.0 + 37.21 + 39.69 + 37.21)$

$= 37.5$ ✓

$Unc = \frac{3.69}{2} = $ ✓✓ So plotting correct X

(ii) $(E_{max})^2 = 2k^2g(h + h_0)^2$

$\therefore k^2v^2 = 2k^2g(h + h_0)^2$ ✓ [subst]

$\therefore v^2 = 2g(h + h_0)^2$

which is right for a falling body if h_0 is some error in measuring the drop height.

(iii)

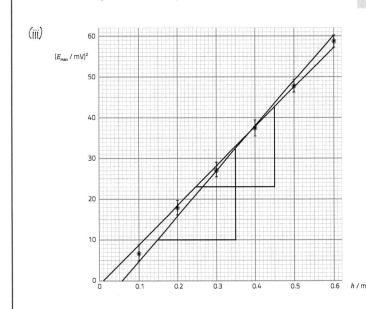

Max gradient $= \frac{33 - 10}{0.35 - 0.15}$ ✓

$= 115 \ mV^2 m^{-1}$ X

Min gradient $= \frac{42 - 23}{0.35 - 0.15}$

$= 95 \ mV^2 m^{-1}$ ✓

Comparing $(E_{max})^2 = 2k^2g(h + h_0)^2$ with

$y = mx + c$, gradient $= 2k^2g$

Max $k = \sqrt{\frac{115}{2 \times 9.81}} = 2.42$ ✓ X [unit]

min $k = \sqrt{\frac{95}{2 \times 9.81}} = 2.20$

$\therefore k = 2.3 \pm 0.1$ ✓ [ecf]

Total   **11 marks /18**

Ffion's answers

(a) (i) The flux linking the coil is changing more quickly as the magnet leaves, because the magnet has gained speed. ✓

MARKER NOTE
Ffion has shown clear understanding.
1 mark

(ii)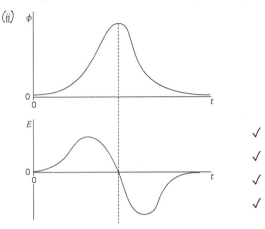

✓
✓
✓
✓

MARKER NOTE
Ffion's includes all four points mentioned in the mark scheme. Ideally, the peaks (positive and negative) should occur when the flux linkage is changing most rapidly (i.e. the points of inflexion) but the marking scheme does not demand that subtle point (the question is difficult enough!).
4 marks

(b) (i) $\overline{\varepsilon_{max}} = \frac{1}{4}(6.0 + 6.1 + 6.3 + 6.1)$ mV $= 6.13$ mV

$\therefore \overline{\varepsilon_{max}^2} = 38 \times 10^{-6}$ V^2 ✓

% uncertainty in $\varepsilon_{max} = \frac{0.15}{6.13} \times 100 = 2.45$ ✓

So % uncertainty in $\varepsilon_{max}^2 = 4.9$

So uncertainty in $\varepsilon_{max}^2 = 4.9 \times 37.6 \times 10^{-6}$ V^2

$\qquad = 1.8 \times 10^{-6}$ V^2 ✓

Point and error bar correct. ✗ [d.p. error]

MARKER NOTE
Ffion's method is not the simplest but is perfectly valid. She loses the last mark because she gives $\overline{\varepsilon_{max}^2}$ to the nearest whole number $\times 10^{-6}$, but the uncertainty to the nearest 0.1×10^{-6}.
3 marks

(ii) Max emf is when magnet is leaving coil and has fallen more than h.

If it has fallen by $(h + h_0)$ ✓

then by energy conservation

$\frac{1}{2}mv^2 = mg(h + h_0)$ ✓

But $\varepsilon_{max} = kv$ ✓ $\therefore \overline{\varepsilon_{max}^2} = 2k^2g(h + h_0)$

MARKER NOTE
The explanation is clear and correct. Ideally, the examiner would have liked to see the cancellation by m in the penultimate line but Ffion clearly knows what she is doing.
3 marks

(iii)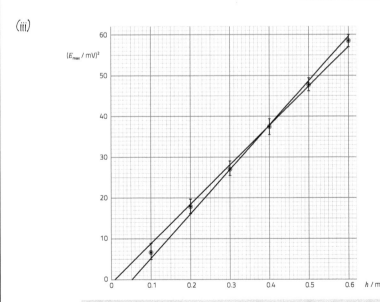

For steepest line, ✓✓

Grad $= \frac{60.0 - 0}{0.60 - 0.05} = 109 \times 10^{-6}$ V^2 m^{-1}

For least steep line, ✓

Grad $= \frac{57.0 - 0}{0.600 - 0.010} = 97 \times 10^{-6}$ V^2 m^{-1}

So gradient $= 103 \pm 6 \times 10^{-6}$ V^2 m^{-1}

So $k = \sqrt{\dfrac{grad}{2g}}$ ✓ $= 2.29 \dfrac{mV}{m\,s^{-1}}$ ✓

Uncertainty $= 2.3 \times \dfrac{6}{103} \dfrac{mV}{m\,s^{-1}}$

$\qquad = 0.13 \dfrac{mV}{m\,s^{-1}}$ ✓

MARKER NOTE
Ffion's lines are the steepest and least steep allowed by the error bars and, although she has not drawn triangles, her gradient calculations show that she has used the largest possible. She has calculated k and its uncertainty correctly, and their units are correct though not given in the standard format. This really is an excellent answer from a candidate who knows what she is about.
6 marks

Total **17 marks /18**

Option A: Alternating currents

Topic summary

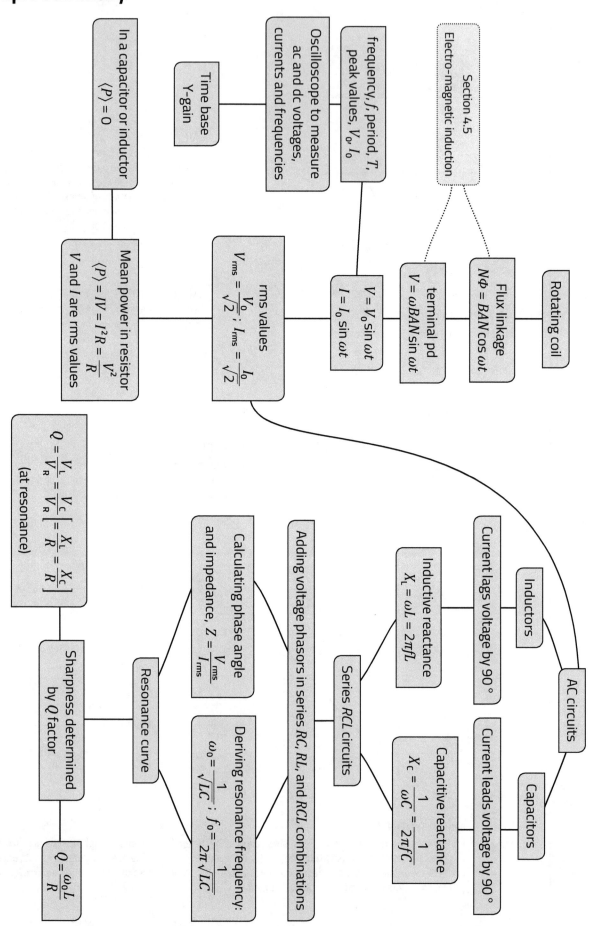

Q1 A square coil, PQRS, of 50 turns, measuring 4.00 cm × 4.00 cm, is turning in a magnetic field of 0.150 T at 20.0 revolutions per second, about an axis passing through the midpoints of PQ and RS. The diagram shows the coil, edge-on, at time $t = 0$.

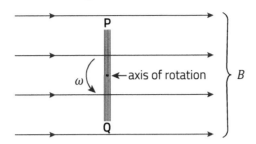

(a) (i) State one time at which the flux linkage is a maximum. .... [1]

(ii) Calculate the maximum flux linkage. [2]

....

....

(b) (i) State one time at which the induced emf is a maximum. .... [1]

(ii) Calculate the maximum emf. [2]

....

....

....

(c) (i) Calculate, to 3 significant figures, the *change* in flux linkage occurring between $t = 0.0115$ s and $t = 0.0135$ s. [3]

....

....

....

....

....

(ii) **Hence** calculate the mean emf induced in the coil during this time interval. [2]

....

....

....

(iii) Compare your answer to (c) (ii) with the instantaneous emf at $t = 0.0125$ s and explain whether or not the comparison is as expected. [3]

....

....

....

....

....

Q2 (a) The coil of a simple ac generator has a resistance of 2.4 Ω and is connected (by slip rings and brushes of negligible resistance) to a 5.6 Ω 'load' resistor. The power dissipated in the **resistor** is 0.30 W.

(i) Calculate the rms pd across the **resistor**. [2]

....

....

....

(ii) Show that the emf of the generator is roughly 2 V rms. [Hint: the resistance of its coil is its *internal resistance*.] [2]

....

....

....

(b) Calculate the rotation frequency (number of revolutions per second) needed to generate this rms emf. The coil is square, measures 5.0 cm × 5.0 cm, consists of 100 turns and rotates in a uniform magnetic field of 0.30 T. [3]

....

....

....

....

....

Q3 A resistor, a capacitor and an inductor are connected in series across a signal generator which is set to 500 Hz. The rms pds across the components are shown in the diagram.

(a) Draw a voltage phasor diagram and use it to determine the rms pd across the terminals of the signal generator. [3]

$f = 500$ Hz

R   C   L

20 V   25 V   15 V

....

....

....

....

....

(b) Ciaran says that the resonance frequency of the circuit is greater than 500 Hz. Discuss briefly whether he is correct. [2]

....

....

....

Q4 An alternating pd is applied across the Y-input terminals of an oscilloscope with its Y-gain set to 200 mV / div and its time base to 2 ms / div. The display is shown:

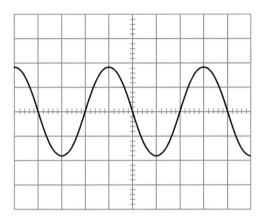

(a) Determine the frequency of the pd. [2]

....

....

....

(b) (i) Determine the rms value of the pd **and** give a reasoned estimate of its absolute uncertainty. [4]

....

....

....

....

....

....

(ii) The following Y-gains were also available: 50 mV / div, 100 mV / div, 500 mV / div. Evaluate whether or not 200 mV / div was the best Y-gain to have chosen for examining this pd. [2]

....

....

....

Q5 A sinusoidal alternating current of peak value 0.15 A is in a series combination of a 12 Ω resistor and a capacitor of reactance 20 Ω. Calculate:

(a) the rms pd across each component; [2]

....

....

(b) the mean power dissipated. [2]

....

....

....

Q6 The graph shows the pd, V, applied across a capacitor of capacitance 0.60 mF:

(a) Show that the peak current is approximately 2 mA. [3]

....

....

....

....

....

(b) **Sketch a graph** of current against time on the grid provided. [2]

(c) Basing your answer on the definitions of *current* and *capacitance*, explain why the maxima of current occur at the times where you have shown them. [3]

....

....

....

....

....

....

Q7 The coil of a small generator has 240 turns and an area of 20 cm². It rotates at 1500 revolutions per minute at right angles to a uniform field of 45 mT. Assuming that the coil has negligible resistance, calculate the mean power it delivers to a load of resistance 120 Ω. [4]

....

....

....

....

....

....

Q8 (a) State one way in which *reactance* is similar to *resistance*. [1]

....

....

(b) State one *way* in which *reactance* differs from *resistance*. [1]

....

....

Q9 The reactances of a capacitor and an inductor at 50 Hz are shown as points marked X_C and X_L on the grid:

(a) **Add lines**, straight or curved as appropriate, to the grid to show how these reactances vary with frequency. [3]

(b) Calculate:

(i) the capacitance of the capacitor; [2]

....

....

....

(ii) the inductance of the inductor.   [2]

....

....

....

Q10   An inductor has a reactance of 15 Ω at 0.5 kHz. This is indicated on the grid.

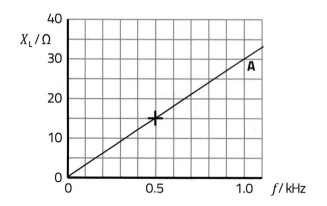

(a) State the significance of graph **A** on the grid.   [1]

....

....

(b) The inductor is connected in series with a resistor of resistance 10 Ω. Sketch a second graph to show
     the variation of the impedance of the combination between 0 and 1.0 kHz.   [3]

     Space for calculations:

Q11   The diagram shows a capacitor and a resistor connected in series across a sinusoidal power supply:

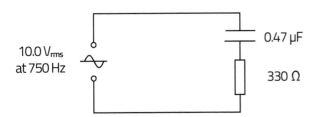

10.0 V$_{rms}$
at 750 Hz

0.47 µF

330 Ω

(a)   (i)   In the space to the right of the circuit diagram, sketch a labelled phasor diagram to show how the rms pds across the capacitor and resistor are related to the rms pd of the supply.   [2]

(ii)   Show that the rms current is roughly 20 mA.   [3]

(b)   (i)   Calculate the rms pd, V_c, across the capacitor.   [2]

(ii)   Determine the phase angle between V_c and the pd applied across the RC combination.   [2]

(iii)   Andrew claims that the rms pd across the resistor is (10.0 V − V_c). Evaluate his claim.   [2]

(c)   Calculate:

(i)   the energy dissipated in the circuit **per cycle**;   [2]

(ii)   the mean energy stored in the capacitor.   [2]

Q12 A coil of wire behaves as an inductance, L, in series with a resistance, R. A group of students uses the circuit shown to investigate how the impedance of a coil varies with frequency.

signal generator

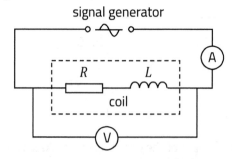

(a) Describe briefly how the circuit can be used to determine the *impedance* of the coil at 100 Hz. The meters are calibrated to read rms values. [2]

....

....

....

(b) With the aid of a labelled phasor diagram, derive the equation:

$$Z^2 = 4\pi^2 L^2 f^2 + R^2$$

[You may use the equation $X_L = \omega L$.] [2]

....

....

....

....

....

....

(c) The students plot a graph of Z^2 against f^2:

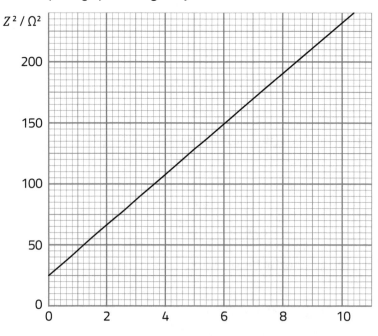

Determine:

(i)　the coil's resistance; [2]

....

....

(ii)　the coil's inductance. [3]

....

....

....

Q13 (a) The values for the resistance, R, and inductance, L, of the coil in question 12 are checked by a different method. A 0.100 µF capacitor is connected in series with the coil in the circuit of 12(a). The signal generator output is kept at 5.00 V rms and the frequency varied. A maximum current of 1.00 A rms is found to occur at 10.6 kHz.

Giving your working, use this information to calculate values for R and L. [3]

....

....

....

....

(b)　Calculate:

(i)　the rms pd across the capacitor at 10.6 kHz; [2]

....

....

(ii)　the Q factor of the circuit. [1]

....

....

Q14 A student reads that the inductance of an air-cored solenoid is given by $L = \mu_0 \dfrac{N^2 A}{\ell}$,where N is the number of turns, A the cross-sectional area and ℓ the length. She winds a 15 mm long, 3.0 mm radius inductor with 25 turns of wire. Calculate the capacitance of the capacitor she needs to connect in series to produce a circuit of resonance frequency 1.6 MHz. [3]

....

....

....

....

Q15 (a) A coil of wire of resistance 33 Ω and inductance 0.070 H is connected to a signal generator set to 100 Hz.

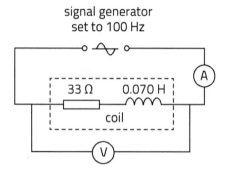

signal generator
set to 100 Hz

33 Ω   0.070 H

coil

The rms current through the coil is 0.20 A. Giving a labelled phasor diagram, show that the rms pd across the signal generator's terminals is roughly 10 V. [5]

(b) (i) Calculate the value of capacitor that must be included in the circuit, in series with the coil, in order to produce resonance at that frequency. [2]

(ii) Calculate the new rms current, assuming that the rms pd across the signal generator's terminals is unchanged. [2]

(c) If the frequency of the signal generator is varied, keeping its rms pd constant, a resonance curve (current against frequency) can be plotted for the set-up in (b). Ciaran claims that the curve will be identical if the experiment is repeated with the 33 Ω resistor replaced by a 66 Ω resistor and the pd of the signal generator doubled. Evaluate to what extent he is correct. [3]

Question and mock answer analysis

Q&A 1   This question is about the circuit shown below. The signal generator is set to give an output of 6.00 V rms at any frequency selected.

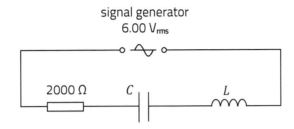

signal generator
6.00 V$_{rms}$

2000 Ω   C   L

(a) It is known that at 65 Hz the reactance of the capacitor is 8.40 kΩ and the reactance of the inductor is 2.10 kΩ.

    (i) Calculate the inductance of the inductor.   [2]

    (ii) (I) Sketch a labelled phasor diagram of voltages at 65 Hz, and show that the rms current is approximately 1 mA.   [5]

        (II) Explain why the rms current has the same value at 260 Hz.   [2]

    (iii) A grid for a resonance curve (rms current against frequency) is given.

        (I) By making appropriate calculations, show that the point already plotted is at the correct position for the **top** of the curve.   [3]

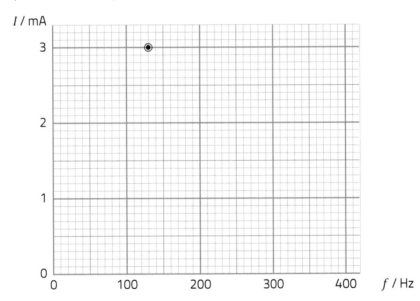

        (II) Use (a)(ii)(I) and (II) to plot two more points, and sketch the resonance curve between $f = 0$ and $f = 400$ Hz.   [2]

    (iv) Calculate the Q factor for the circuit.   [2]

(b) The inductor is now replaced by one of inductance $\frac{1}{2}L$, and the capacitor by one of capacitance $2C$. Emma claims that the resonance frequency and the current at resonance will stay the same, but the curve will be sharper. Evaluate her claims.   [4]

What is being asked

The question opens in a non-frightening way, with (a)(i) just involving the use of a standard equation. In part (ii) the examiner wishes to test understanding of phasor diagrams and has saved candidates the trouble of working out reactances from L and C values! There is more than one way of proceeding in the second part, but the low mark allocation suggests that this part can be done simply. Although candidates are being asked about a point on a graph in (iii)(I), very standard calculations are being tested but part (II) requires recall of knowledge.

(a)(iv) is a straightforward test of the candidate's ability to calculate a specific quantity, but perhaps its inclusion at this point could be a clue to help with the next part ... (b) in which the candidate has to evaluate three claims, choosing the order in which to tackle them, and the strategy – hence the AO3 classification.

Mark scheme

Question part			Description	AOs			Total	Skills	
				1	**2**	**3**		**M**	**P**
(a)	(i)		$L = \dfrac{X_L}{\omega}$ (transposition at any stage) [1] $L = 5.1$ H [1]		2		2	2	
	(ii)	(I)	Pattern of 3 lines or arrows labelled to identify as X_L, R, X_C or V_L, V_R, V_C even if X_L, X_C or V_L, V_C wrong way round. [1] X_L (or V_L) shown π in advance of X_C (or V_C) [1] $Z = \sqrt{(8400 - 2100)^2 + 2000^2}$ equiv or by impl [1] $I = V / Z$ used [1] $I = 0.91$ mA [1]	1 1 1 1	1		5	3	
		(II)	Values of X_L and X_C swap over or equivalent [1] $(8400 - 2100)^2 = (2100 - 8400)^2$ or equiv [1]		1 1		2	2	
	(iii)	(I)	X_L shown to equal X_C e.g. both stated to be 4.2 kΩ, at 130 Hz [1] $I = 6.00$ [V] / 2000 [Ω] = 3.0 mA as plotted [1] Resistance only, or cancelling of reactances, shows current has greatest value. [1]		1 1 1		3	1	
		(II)	Points at 65 Hz and 260 Hz plotted correctly and bell-shaped curve drawn through the 3 points [1] Curve goes through origin and looks as if it *could* be asymptotic to f axis for large f. [1]	1	1		2		
	(iv)		Use of $X = 4.2$ kΩ [1] $Q = 2.1$ [1]	2			2		
(b)			$\omega_0^2 = \dfrac{1}{\frac{1}{2}L2C} = \dfrac{1}{LC}$, or equiv argument [1] [V (rms) and] R the same so I same at res [1] X_L at resonance halved because ω_0 the same and L halved or equivalent argument [1] Therefore Q halved **and** curve *less* sharp so Emma incorrect. [1]			4	4		
Total				**7**	**9**	**4**	**20**	**8**	**0**

Rhodri's answers

(a)(i) $L = \dfrac{2100}{65} = 32\ H$

MARKER NOTE
Rhodri has divided the reactance by the frequency, f, rather than by ω. This has cost him both marks because he has not shown any intention to divide by ω.
0 marks

(ii)(I)

8400

2000

2100

✓ X

$Z = \sqrt{(8400 - 2100)^2 + 2000^2}$ ✓

$= 5974$ X

$I = \dfrac{6.00}{5974}$ ✓ ecf

$= 1.00 \times 10^{-3}\ A$ X

$= 1.00\ m$

MARKER NOTE
Rhodri's labelling is minimal, but enough to identify the phasors – and to show that he has the inductive and capacitative phasors the wrong way round. He uses the correct equation for Z, and puts in the data correctly, but has made some slip in evaluation (which is only penalised once). He uses $I = V / Z$ correctly.
3 marks

(II) Because the current reaches a peak and comes back down, there must be another frequency at which the current has the same value as at 65 Hz.

MARKER NOTE
Rhodri has not dealt with the specific frequency of 260 Hz. He has either misunderstood the question or has been unable to make a proper attempt.
0 marks

(iii) (I) At top of curve, Z has lowest value possible ✓
so $Z = R$, so $I = 6.00 / 2000 = 3 \times 10^{-3}\ A = 3\ mA$, so point is correctly plotted ✓

MARKER NOTE
Again Rhodri has not considered the actual frequency, but he has explained correctly why 3.0 mA is the greatest current.
2 marks

(II).

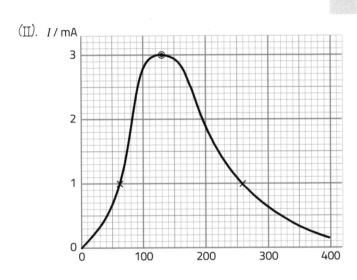

MARKER NOTE
Rhodri has drawn a very convincing resonance curve of current against frequency.
2 marks

(iv) $Q = \dfrac{X_L}{R} = \dfrac{2100}{2000} = 1.1$ X

MARKER NOTE
Rhodri has not realised that $Q = X_L/R$ must be evaluated using X_L at resonance.
0 marks

(b) At resonance $\omega_0 L = \dfrac{1}{\omega_0 C}$ $\therefore \omega_0^2 = \dfrac{1}{LC} = \dfrac{1}{(\frac{1}{2}L)\,2C}$ ✓
Therefore resonance frequency the same.
Resistance the same so current the same. ✓

Sharpness depends on resistance, so sharpness the same.
Emma has got this one wrong.

MARKER NOTE
Rhodri has shown clearly that Emma's first two claims are correct. However he has not appreciated the role of inductance, and in particular, Q, in determining sharpness. This has led to the wrong analysis of Emma's mistake.
2 marks

Total   **9 marks /20**

Ffion's answers

(a)(i) $X_L = \omega L$

$\therefore L = \dfrac{X_L}{\omega}$ ✓ $= \dfrac{2100}{2\pi \times 65} = 16$ H ✗

(ii)(I)

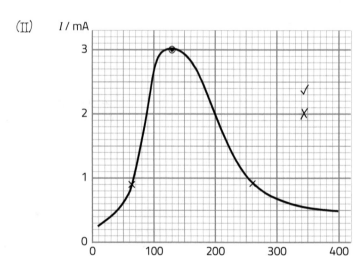

$V = \sqrt{I^2(X_L - X_C)^2 + I^2 R^2}$

$I = \dfrac{V}{\sqrt{(X_L - X_C)^2 + R^2}}$ ✓

$= \dfrac{6.00}{\sqrt{(2100 - 8400)^2 + 2000^2}}$ ✓ $= 0.908$ mA ✓

(II) $X_C \propto \dfrac{1}{f}$ and $260 = 4 \times 65$

$\therefore$ at 260 Hz, $X_C = \frac{1}{4}$ 8.4 kW = 2.1 kW

Also $X_L \propto f \therefore$ at 260 Hz, $X_L = 4 \times 2.1$ kW

So X_L and X_C are simply exchanged. ✓

(iii)(I) At 130 Hz, $X_C = 4.2$ kW, $X_L = 4.2$ kW ✓

Reactances cancel so $Z = R$ and is a minimum.

Therefore $I = 6.00/2000 = 3.00$ mA ✓ and is

a maximum at 130 Hz. ✓

(II)

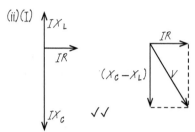

(iv) $Q = \dfrac{X_L}{R} = \dfrac{2100}{2000}$ ✓ $= 2.1$ ✓

(b) $X_C = \dfrac{1}{C}$ so doubling C halves X_C for any frequency.

$X_L = \omega L$ so halving L halves X_L. Therefore we still have

$X_C = X_L$ at 130 Hz. ✓

Resistance unchanged so current still 3.0 mA ✓

Q is decreased, so curve is now

less sharp – and Emma is wrong!

[not enough]

Total   **15 marks /20**

Option B: Medical physics

Topic Summary

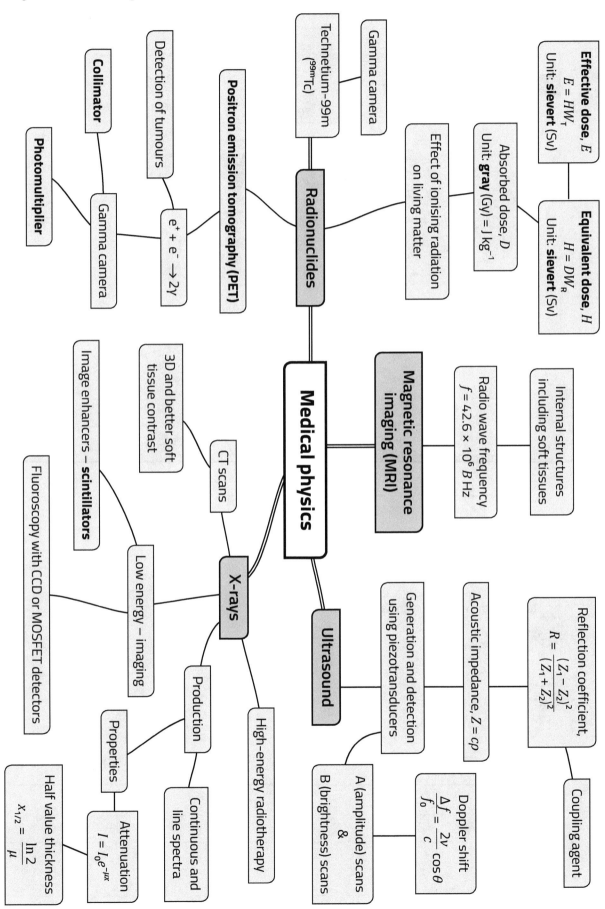

Q1 State some of the properties of X-rays that make them suitable for imaging bones. [3]

....

....

....

....

Q2 An X-ray tube is shown:

(a) Explain the process by which the continuous spectrum is produced. [2]

....

....

....

....

(b) Explain the process by which the line spectrum is produced. [2]

....

....

....

....

(c) Explain why a vacuum is required. [1]

....

....

....

....

(d) Calculate the speed of the electrons just before they strike the tungsten target and comment on the validity of your calculation. [3]

....

....

....

....

(e) Calculate the minimum wavelength of X-rays emitted. [2]

....

....

....

....

(f) The tube current is 14.5 mA and 5.1 W of X-ray power is produced.

(i) Calculate the efficiency of the X-ray tube. [2]

....

....

....

....

(ii) Explain why water-cooling of the tungsten target is required. [2]

....

....

....

....

(iii) A book on X-rays states that a good approximation for the efficiency of an X-ray tube is:

Efficiency of X-ray tube (%) = pd (in kV) × atomic number of target × 10^{-4}

Determine whether this equation is a good approximation for this X-ray tube. The atomic number of tungsten is 74. [2]

....

....

....

....

Q3 The X-ray spectrum of an X-ray tube operating at 60 kV is shown.
On the same diagram, sketch the X-ray spectrum for the same tube operating at 30 kV. [3]

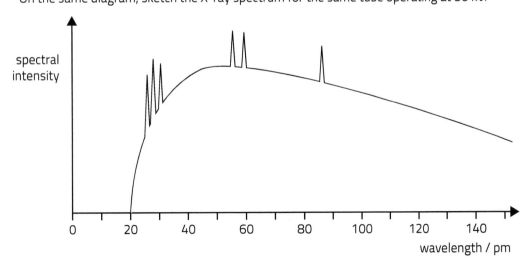

Q4 A book on medical physics states:

A piezoelectric transducer emits ultrasound by using the reverse piezoelectric effect while a piezoelectric detector uses the piezoelectric effect.

(a) (i) Explain the difference between the *reverse piezoelectric effect* and the *piezoelectric effect*. [2]

....

....

....

....

(ii) Explain how a piezoelectric transducer can be used to send short ultrasound pulses. [2]

....

....

....

....

(b) Explain the difference between an ultrasound A-scan and an ultrasound B-scan. [4]

....

....

....

....

....

....

(c) The table shows values of density, speed of sound and acoustic impedance for soft tissue and bone:

Body tissue	Density / kg m^{-3}	Speed of sound / m s^{-1}	Acoustic impedance / ....
Soft tissue		1590	1.70×10^6
Bone	1650		6.73×10^6

(i) Complete the table including the correct unit for acoustic impedance. [3]

(ii) Show that the reflection coefficient between soft tissue and bone is 36%. [1]

....

....

....

....

(iii) A new-born baby's brain can be imaged using an ultrasound scan because there is no calcification of the skull. A radiologist states that the detected reflections from inside the brain will be many times weaker for an adult compared with that of a new-born child. She further states that the intensity of reflections will be 13% of those for the baby because $0.36^2 = 0.13$, for the reflection on the way in and on the way out. Determine to what extent the radiologist is right and correct her figures (if they need correcting). [3]

Q5 (a) The half thickness $(x_{\frac{1}{2}})$ is defined as the thickness of a material that reduces the intensity of X-rays to half its original value. Show that the half thickness is related to the absorption coefficient (μ) by the following relationship:

$$\mu x_{\frac{1}{2}} = \ln 2$$ [3]

(b) In muscle fibre, the half thickness for certain X-rays is 3.7 cm. Calculate the percentage reduction in X-ray intensity when the beam of X-rays passes through 5.0 cm of muscle fibre. [3]

Q6 (a) An ultrasound technician takes images of a foetus inside the womb using an ultrasound B-scan. Explain briefly how an ultrasound B-scan image is produced and why a coupling gel is crucial in obtaining the image. [4]

(b) (i) As part of the ultrasound scan, the maximum speed of blood flow in the aorta of the foetus is measured. Explain how the blood flow can be measured using the Doppler shift. [3]

....

....

....

....

....

(ii) Ultrasound of frequency 5.50 MHz is used, the ultrasound enters the aorta at an angle of 5.0° to the direction of blood flow and the speed of sound in blood is 1580 m s^{-1}. Calculate the change in frequency of the ultrasound when the blood flow is measured as 105.8 cm s^{-1}. [2]

....

....

....

....

Q7 A simplified diagram of a fluoroscopy set-up is shown.

(a) Explain the purpose of the anti-scatter grid and the scintillator screen. [2]

CCD video camera
scintillator screen
anti-scatter grid
patient
X-ray filter
simple collimator
X-rays
X-ray tube

....

....

....

....

....

....

(b) Explain why the design of the scintillator screen is important in decreasing the risk of cancer to the patient. [2]

....

....

....

....

....

Q8   High energy X-rays (around 10 MeV) are used in radiation therapy.

(a)   Explain why higher energy X-rays are required for radiation therapy than for X-ray imaging.   [2]

....

....

....

(b)   Explain why higher intensity beams are required for radiation therapy.   [2]

....

....

....

....

Q9   (a) Explain the role of radio waves in magnetic resonance imaging (MRI).   [3]

....

....

....

....

....

(b)   Calculate the range of frequency of radio waves used by an MRI machine which has a magnetic field strength varying from 1.53 T to 1.94 T.   [2]

....

....

....

(c)   An international netball player has a problem with her knee joint. An orthopaedic surgeon recommends an MRI scan, traditional X-ray images and an ultrasound B-scan to diagnose the problem. Evaluate the strengths and weaknesses of each of these techniques in imaging the knee joint.   [6]

....

....

....

....

....

....

....

....

....

Q10 (a) In a PET scan, two gamma rays are detected by two gamma detectors **A** and **B**. The gamma ray is detected by detector **A** 237 ps before the other gamma ray is detected by detector **B**. Place a cross on the diagram at the location of the source of the two gamma rays. [3]

detector A

detector B

0cm 1 2 3 4 5 6 7 8 9 10 11 12 13 14 15 16 17 18 19 20 21 22 23 24 25 26 27 28 29 30

[Space for calculations]

(b) Explain briefly the physical principles of a PET scan. [4]

....

....

....

....

....

....

....

....

Q11 Explain briefly how a CT scanner can obtain a 3D image of a patient's innards. [4]

....

....

....

....

....

....

....

....

Q12 Tables of the radiation weighting factor and tissue weighting factor are shown:

Radiation type and energy range	Radiation weighting factor W_R
X-rays and γ-rays, all energies	1
Electrons, positrons, muons, all energies	1
Neutrons:	
<10 keV	5
10 keV to 100 keV	10
>100keV to 2 MeV	20
>2 MeV to 20 MeV	10
>20 MeV	5
Protons	2 to 5
α-particles	20

Tissue	Tissue weighting factor W_T
Red bone-marrow, colon, lung, stomach, breast, other tissues	0.12
Gonads	0.08
Bladder, oesophagus, liver, thyroid	0.04
Bone, brain, salivary glands, skin	0.01

(a) A patient is irradiated uniformly with a beam of 1 MeV neutrons. The equivalent dose received by the liver is 550 mSv.

(i) Calculate the contribution of the liver to the patient's effective dose. [2]

....

....

....

(ii) The neutrons are absorbed uniformly by the patient's whole body and the patient has a mass of 94 kg. Calculate the total energy of the neutrons absorbed. [3]

....

....

....

....

....

(b) A radiologist states that breathing or swallowing an α-particle source may be the most dangerous thing you could possibly do with ionising radiation. Use data from the tables to evaluate this statement. [3]

....

....

....

....

....

Q13 The diagram shows a patient who has received the radioactive tracer technetium-99m. A gamma camera is then used to image the patient's kidneys.

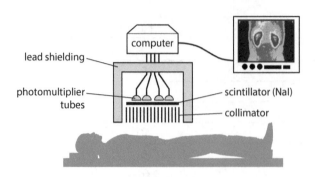

(a) Explain the purpose of the scintillator, collimator and photomultiplier tubes in the gamma camera. [3]

....

....

....

....

....

....

(b) State some properties you would expect of technetium-99m for it to be a suitable radionuclide to be used with the gamma camera. [3]

....

....

....

....

....

....

Question and mock answer analysis

Q&A 1

(a) Calculate the operating pd of the X-ray tube whose spectrum is shown.   [3]

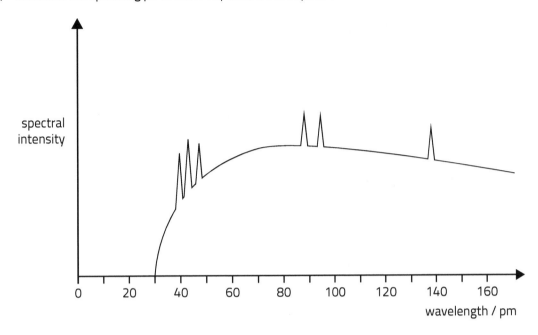

(b) On the diagram above, sketch the X-ray spectrum for the same tube when the operating pd is increased by 50%.   [3]

(c) Five years after an operation for a replacement heart valve, a patient is required to have an annual scan to check the performance of the valve. Evaluate the suitability of the following techniques in carrying out the annual scan: [5]

Fluoroscopy   CT   MRI   Ultrasound   PET

What the question is asking

Part (a) are AO2 marks. They require the candidate to obtain a value for the minimum wavelength from the graph and then use this to obtain the accelerating pd of the X-ray tube.

Part (b) are mainly AO1 marks but there is one AO2 mark for calculating the new minimum wavelength. The two AO1 marks are for knowing that the basic shape of the spectrum will remain the same and for knowing that the position of the spikes must remain at the same wavelength.

Part (c) are all AO3 marks. The candidates must evaluate the suitability of each of the five imaging techniques in obtaining information about a heart valve. The strengths and weaknesses of each technique are required for a good answer.

Mark scheme

Question part			Description	AOs			Total	Skills	
				1	2	3		M	P
(a)			Minimum wavelength read correctly (30 pm) [1] $eV = hc/\lambda$ applied [1] Correct answer = 41.4 kV [1]		3		3	1	
(b)			Minimum wavelength = 20 pm (written or in graph) [1] All peaks in same place [1] Background spectrum following same pattern [1]	1 1	1		3	1	
(c)			Any 2 points for each method – 1 mark **Fluoroscopy** – moving images, flow can be checked from videos, X-ray dosage, expensive. **CT** – image only (not moving), X-ray dosage, blood flow can be checked (using contrast media), expense **MRI** – good images, expensive, non-ionising, blood flow can be checked (using contrast/real time MRI) **Ultrasound** – B-scan cheap, moving images, non-ionising, Doppler gives blood flow too (best method) **PET** – radiation dosage, low resolution, still images, not useful for flow (expensive or availability not a mark option because it will not work)			5	5		
Total				2	4	5	11	2	0

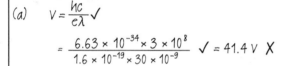

Rhodri's answers

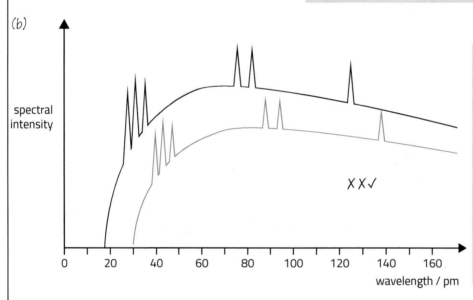

(a) $V = \dfrac{hc}{e\lambda}$ ✓

$= \dfrac{6.63 \times 10^{-34} \times 3 \times 10^{8}}{1.6 \times 10^{-19} \times 30 \times 10^{-9}}$ ✓ = 41.4 V ✗

MARKER NOTE
Rhodri's answer is perfect except that he has made a mistake with the power of ten (pm is 10^{-12} m not 10^{-9} m).

2 marks

(b)

spectral intensity

X X ✓

wavelength / pm

MARKER NOTE
Rhodri's answer looks ok but there are two important mistakes. First, all the peaks have moved to lower wavelengths. Second, the minimum wavelength should be 20 pm (increasing by 50% is increasing by a factor of 1.5, which means that the wavelength decreases by a factor of 1.5). He only gains the mark for the background spectrum.

1 mark

(c)   Fluoroscopy is expensive and useless because you don't want to put radioactive tracers inside the patient every year ✗. CTs give good images of the heart but has ionising radiation ✓. MRIs are expensive but don't have ionising radiation ✓.

Ultrasound is useless because the resolution isn't good enough ✗. PET scans are also low resolution but are far too expensive and not available in all hospitals. Overall, MRI would be best (unless the patient has a pace-maker).

MARKER NOTE
Rhodri has completely misunderstood fluoroscopy which would use iodine as a contrast medium but certainly has no radioactive tracer. He makes two acceptable points about CT scans (good image and ionising radiation). He also makes two acceptable points about MRI scans. He requires a little more to obtain the PET scan mark).

2 marks

Total   **5 marks /11**

Ffion's answers

(a)   Photon energy = $\frac{hc}{\lambda}$

= 6.63×10^{-15} J ✓

= 41.4 keV ✓

So V = 41.4 kV ✓

MARKER NOTE

Ffion's answer is set out differently from the mark scheme. She first calculated the photon energy in J, converted this to eV which will then be the same (numerically) as the pd.

3 marks

(b)

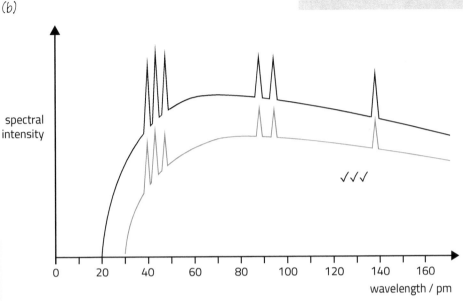

MARKER NOTE

Ffion's answer is awarded all 3 marks because she meets the criteria for all the individual marks. The only mark in doubt is the background radiation (this background continues underneath the peaks). The examiner has allowed it on this occasion because it is easier to draw the background spectrum first and add the peaks later.

3 marks

(c)   The best and most cost-effective method would be to use non-ionising ultrasound B-scans (maybe even transoesophageal echocardiogram). These would provide cheap moving images and a Doppler ultrasound scan would give details of blood-flow too ✓. MRI is too expensive and all the other techniques involve ionising radiation ✓. Also, ultrasound is the only technique that will provide detailed blood-flow values and this is essential to check the operation of the replacement valve. Consultation over, my fee is $4000, my secretary accepts all major credit cards, thank you!! ☺ CT ✓, PET ✓

MARKER NOTE

Ffion has detailed knowledge about ultrasound scans and thoroughly deserves that mark. She has also identified that MRI is expensive and does not involve ionising radiation (this is implied but not stated). Because she has identified all the ionising and non-ionising radiations, she has made 1 good point about fluoroscopy, CT and PET but she goes further and states that none of these methods will give detailed blood-flow values. This is usually true of CT scans and definitely true of PET scans but not true of fluoroscopy where directions and speeds of flow can be calculated from the videos. Overall, she has 2 points about CT, PET, ultrasound and MRI but only 1 correct on fluoroscopy. Her final sentence, although amusing, will not be viewed favourably by examiners.

4 marks

Total	10 marks /11

Option C: Physics of sport

Topic summary

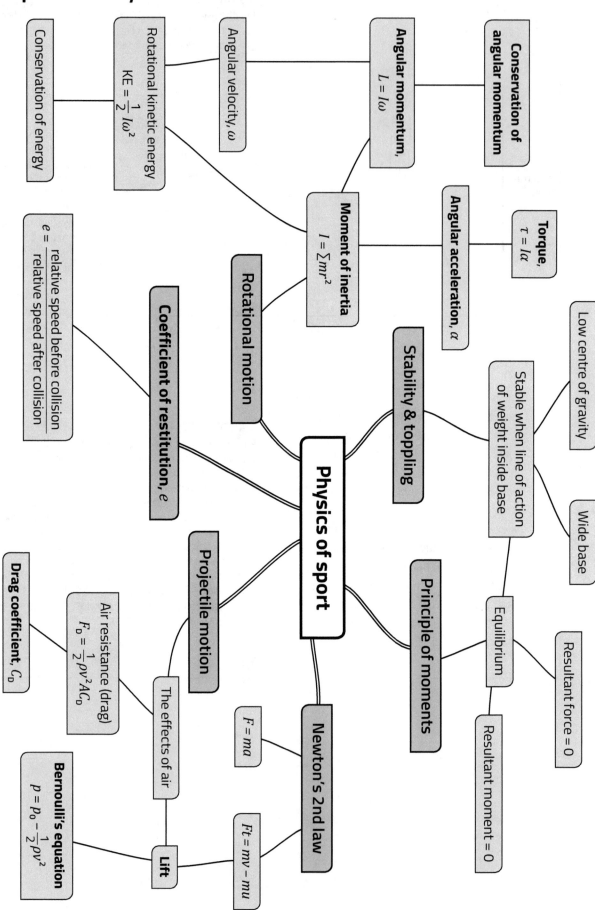

Conservation of energy

Rotational kinetic energy
$$KE = \frac{1}{2}I\omega^2$$

Angular velocity, ω

Angular momentum,
$L = I\omega$

Conservation of angular momentum

Moment of inertia
$$I = \sum mr^2$$

Angular acceleration, α

Torque,
$\tau = I\alpha$

$e = \dfrac{\text{relative speed before collision}}{\text{relative speed after collision}}$

Coefficient of restitution, e

Rotational motion

Stability & toppling

Low centre of gravity

Stable when line of action of weight inside base

Wide base

Physics of sport

Principle of moments

Equilibrium

Resultant force = 0

Resultant moment = 0

Projectile motion

Drag coefficient, C_D

Air resistance (drag)
$$F_D = \frac{1}{2}\rho v^2 A C_D$$

The effects of air

Newton's 2nd law

$F = ma$

$Ft = mv - mu$

Bernoulli's equation
$$p = p_0 - \frac{1}{2}\rho v^2$$

Lift

Q1 It is sometimes said that short, stocky players have a great advantage when playing football. Explain why this might be true, using stability. A simple diagram might help your answer. [3]

....

....

....

....

....

....

Q2 An athlete carries out a leg raise exercise as shown. The simplified diagram below shows the weight of the leg, the weight of the foot and the force (F) exerted on the leg by the relevant muscle. Calculate the force, F, required to hold the leg in equilibrium. [3]

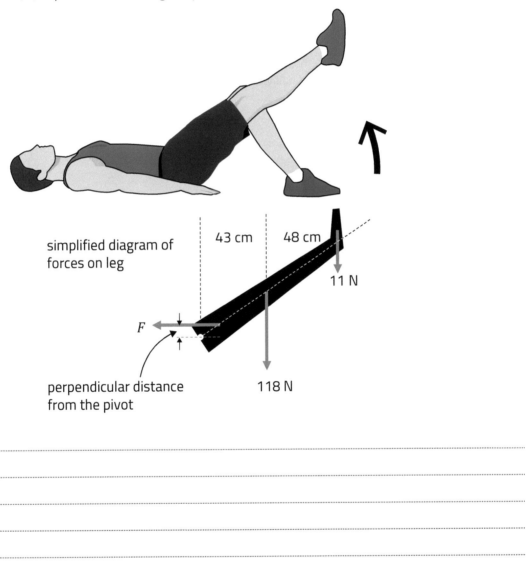

simplified diagram of forces on leg

43 cm   48 cm

11 N

F

perpendicular distance from the pivot

118 N

....

....

....

....

....

Q3 (a) A thin, hollow cylinder is rotated about its central axis.

(i) Explain why its kinetic energy (KE) is given by:

$$KE = \tfrac{1}{2}mr^2\omega^2$$

where r is the radius of the hollow cylinder, m its mass and ω its angular velocity. [3]

(ii) Show that, when the hollow cylinder rolls down an inclined plane from rest, half its kinetic energy will be linear KE and half will be rotational KE. [3]

(iii) When the hollow cylinder has rolled down an inclined slope a vertical distance of 0.30 m, its rotational KE is 0.45 J. Calculate the mass of the hollow cylinder. [3]

(b) The hollow cylinder and a snooker ball are both rolled down an inclined plane from the same height. Explain why the snooker ball arrives at the bottom of the inclined plane before the hollow cylinder (moment of inertia of a snooker ball = $\tfrac{2}{5}mr^2$). [3]

Q4 Serena Williams strikes a forehand topspin shot with extreme power. In the collision between racket and ball, the ball changes direction from 32 m s^{-1} due east to 48 m s^{-1} due west in a time of 6.8 ms. The ball also changes spin from 1200 revolutions per minute clockwise to 2550 revolutions per minute anticlockwise. The mass of the tennis ball is 58.2 g and its diameter is 66.9 mm.

(a) Calculate the linear and angular acceleration of the tennis ball. [5]

(b) Calculate the net torque and net force acting on the tennis ball during the collision.
[Moment of inertia of tennis ball = $\frac{2}{5}mr^2$] [4]

(c) Charles states that the rotational kinetic energy of the tennis ball is now greater than its linear KE. Determine whether, or not, he is correct. [3]

(d) Serena strikes the tennis ball at an angle of 6.5° above the horizontal from a height of 0.95 m above the ground. If the ball travels further than 31.2 m it will land beyond the court. Determine whether, or not, Serena has hit the ball too far.
[Ignore any effects of spin or air resistance for this part of the question.] [4]

(e) (i) The tennis ball has a drag coefficient of 0.60. Calculate the initial drag force acting on the tennis ball and comment whether, or not, drag can be ignored to a good approximation. [4]

(ii) The topspin on the tennis ball means that its lift is −2.0 N, i.e. there is a downward force of 2.0 N acting on the ball. Discuss how this force and the drag force might change your conclusion to part (d). [3]

Q5 When taking a shower after a match, a rugby player noticed that the shower curtain moved inwards when the shower was turned on. Explain this using the Bernoulli equation. [3]

Q6 The windsurfer shown is in equilibrium. Explain why the windsurfer does not fall backwards into the sea. [3]

Q7 Calculate the coefficient of restitution between snooker balls, using the data in the diagram: [3]

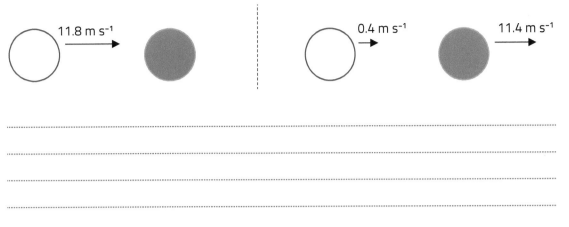

....

....

....

....

....

Q8 A trampolinist performs a triple somersault. She starts the jump with her body in the straight position but her rate of rotation increases when she goes into the tuck position. Explain why her rate of rotation increases when she goes into the tuck position. [3]

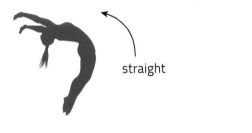
straight

tuck

....

....

....

....

....

Q9 (a) A tennis ball of mass 58 g is struck at a wall with a velocity of 63 m s^{-1} due north. It rebounds off the concrete wall with a velocity of 49 m s^{-1} due south. Calculate the mean force exerted by the wall on the ball given that the ball is in contact with the wall for 6.5 ms. [3]

....

....

....

....

....

(b) (i) Calculate the coefficient of restitution of the tennis ball and the concrete wall. [2]

....

....

....

(ii)   To what fraction of its original height would you expect the tennis ball to rebound when dropped onto a concrete (hard court) floor? [2]

....

....

....

Q10   A golf club is in contact with a golf ball for 257 μs. During this time, a golf ball of mass 45.93 g and diameter 42.67 mm acquires a linear speed of $85\,\mathrm{m\,s^{-1}}$ and a spin rate of 2700 revolutions per minute.

(a)   Calculate the mean resultant force acting on the ball during the collision. [2]

....

....

....

(b)   Calculate the mean angular acceleration of the golf ball. [3]

....

....

....

....

(c)   Calculate the mean tangential force providing the ball's angular acceleration during the collision. [Moment of inertia of a solid sphere = $\frac{2}{5}mr^2$] [4]

....

....

....

....

....

....

(d)   Draw the two components of the resultant force acting on the golf ball to scale in the directions indicated by the dotted lines. [2]

golf club head

(e) Calculate the initial ratio of rotational kinetic energy to linear kinetic energy for the golf ball. [3]

Q11 A football has a diameter of 22.0 cm. The ball is kicked by a goalie so that it moves from ground level at a speed of 18.1 m s^{-1} at an angle of 21.2° above the horizontal.

(a) Show that the ball should travel a horizontal distance of approximately 20 m before hitting the ground, if the effects of air resistance and spin are ignored. [3]

(b) The coefficient of drag of the football is 0.195. Calculate the initial drag force acting on the football. [ρ_{air} = 1.25 kg m^{-3}] [2]

(c) The ball is kicked from left to right with backspin. Explain why the actual distance travelled by the football is likely to be greater than your answer to part (a) [Hint: use the diagram]. [3]

direction of motion of football

(d) Draw labelled arrows to represent the three forces acting on the football (the ball has backspin so that the lift force is positive). [3]

direction of travel

(e) Evaluate the effects of spin and air resistance on the flight time of the ball, its mean horizontal speed and range. [6]

....

....

....

....

....

....

....

....

(f) The spin rate of the ball drops from 2700 revolutions per minute to approximately half this value when the ball lands. Explain whether or not this contradicts the principle of conservation of angular momentum. [2]

....

....

(g) The trajectory of the football in a vacuum is shown by the dotted line. Sketch the expected trajectory of the ball in air with backspin and air resistance (the expected range in air is 25 m). [3]

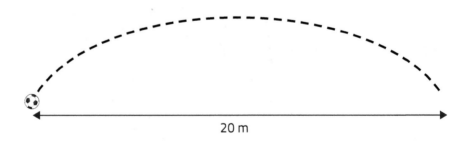

20 m

Question and mock answer analysis

Q&A 1 Data relating to snooker balls can be seen in the table:

Snooker ball data		
Diameter	Mass	Drag coefficient
52.5 mm	165 g	0.48

A stationary snooker ball is struck with a cue tip. The tip is in contact with the ball for 1.15 ms during which time the ball accelerates to $13.6\,\mathrm{m\,s^{-1}}$.

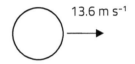

$13.6\ \mathrm{m\ s^{-1}}$

cue tip

(a)   Calculate the resultant force acting on the ball.   [3]

(b)   As the diagram shows, the snooker ball is not struck in its centre, so that a couple of 30.6 N m is exerted on the ball at the same time. Calculate the angular acceleration of the snooker ball.   [3]

(c)   The frictional force exerted by the cloth on the snooker ball is a constant and is given by:

Frictional force = 0.060 × weight of snooker ball

(i)   Calculate the speed at which friction and air resistance are equal for a snooker ball when the density of air is $1.25\ \mathrm{kg\ m^{-3}}$.   [4]

(ii)   Archibald claims that the effect of air resistance is greater than the effect of friction for the motion of a snooker ball. Discuss to what extent Archibald is correct.   [3]

What the question is asking

Part (a) are mainly AO2 skills where Newton's 2nd law must be used to obtain the resultant force. Part (b) is very similar but is to do with rotational motion rather than linear motion. The moment of inertia of the snooker ball will have to be calculated too. Once again, (c)(i) are mainly AO2 skills and the equation for drag must be equated to the frictional force given in the question. Part (c)(ii) are AO3 skills. Friction and air resistance must be compared and then a logical conclusion should be presented

Mark scheme

Question part			Description	AOs			Total	Skills	
				1	2	3		M	P
(a)			Quoting or using Newton's 2nd law in either $F=ma$ or momentum form ($Ft = mv-mu$) [1]	1			3		
			Rearrangement or $\dfrac{0.165 \times 13.6}{0.00115}$ seen [1]		1			1	
			Correct answer = 1950 N [1]		1			1	
(b)			Substitution into moment of inertia equation i.e. $\frac{2}{5} \times 0.165 \times 0.02625^2$ (or 4.13×10^{-5} seen) [1]	1			3		
			$\alpha = \dfrac{\tau}{I}$, i.e. rearrangement (or implied) [1]		1			1	
			Correct answer = 673 000 rad s^{-2} [1]		1			1	
(c)	(i)		Use of drag equation i.e. $F = \frac{1}{2}\rho v^2 A C_D$ [1]	1			4		
			Use of weight equation i.e. $W = mg$ [1]		1				
			Equated the forces i.e. $0.06mg = \frac{1}{2}\rho v^2 A C_D$ [1]		1			1	
			Final answer = 12.2 m s^{-1} [1]		1			1	

(ii)		Drag seems as large or comparable to friction (any comment comparing sizes) [1] Drag will decrease with decreasing speed (or converse) [1] Good final conclusion, e.g. drag more important at high speeds, overall fiction more important because drag drops (quickly), also accept both seem as important as each other (due to similar sizes) [1]				3	3		
Total						3	7	3	13	6	0

Rhodri's answers

(a) Force $= 0.165 \times 13.6$

$= 2.244\,N$ ✗

MARKER NOTE
Rhodri's answer only calculates the change in momentum of the snooker ball. There is no mark for this and he cannot be awarded any marks. Although he only needs to divide this answer by the time, none of the correct steps are present here.

0 marks

(b) Moment of inertia $= \frac{2}{3}mr^2$ ✗

$= 3.03 \times 10^{-4}$

$\alpha = \dfrac{30.6}{3.03 \times 10^{-4}}$ ✓

$= 101000\,\text{rad s}^{-2}$ ✗ no ecf

MARKER NOTE
Rhodri has used the incorrect equation for the moment of inertia (he has used the hollow sphere rather than the solid sphere equation). The method for the 2nd mark is correct.

1 mark

(c) (i) Drag $= \frac{1}{2} 1.25 v^2 (4\pi r^2) C_D$ ✗

Weight $= 9.81 \times 0.165 = 1.62$ ✓

$0.06 \times 1.62 = \frac{1}{2} 1.25 v^2 (4\pi \times 0.0525^2) C_D$ ✓

$v = 3.1\,\text{m s}^{-1}$ ✗ (no ecf)

MARKER NOTE
Rhodri has an excellent effort at this part but slips cost him two marks. He loses the 1st mark because he has used the surface area rather than the cross-sectional area and he has used the diameter instead of the radius. He obtains the 2nd mark for the weight equation and the 3rd for equating the forces. The final answer is wrong and no ecf can be awarded because the candidate has made a mistake in this section and not a previous section.

2 marks

(ii) If the forces are equal at such a low speed ✓(ecf) then it looks as though air resistance is more important than friction and Archie might just be correct. ✓ (ecf)

MARKER NOTE
Rhodri has a good effort at this part too and obtains 2 marks. He compares the forces (by saying they are equal at a low speed) and then comes to a very sensible conclusion (applying ecf for his low value in the previous part). He cannot gain the 2nd mark because there is nothing here relating to how the drag varies with speed.

2 marks

Total   **5 marks /13**

Ffion's answers

(a) Force $= \dfrac{0.165 \times 13.6}{0.115}$ ✓✓

$= 2000 \, N$ ✓ BOD (rounded to 2sf)

MARKER NOTE
Ffion's answer is minimalist but the answer is correct and so is the method so she obtains full marks.

3 marks

(b) $I = \frac{2}{5}mr^2$ ✓ $= 4.55 \times 10^{-5}$

$\alpha = \dfrac{30.6}{4.55 \times 10^{-5}}$ ✓ $= 672850 \, rad \, s^{-2}$ ✓

Must be wrong. Do a quick check.

$w = \alpha t = 774 \, rad \, s^{-1}$

So $f = 774 / 2\pi = 123 \, Hz$

Which seems quick but maybe ok.

MARKER NOTE
Ffion's answer is awarded all 3 marks because she meets the criteria for all the individual marks and the final answer is correct. She is not satisfied by the seemingly enormous angular acceleration and so does a quick check to see what the angular velocity would be (by multiplying the angular acceleration by the time). This was not required but was a very sensible thing to do when confronted by such a large number.

3 marks

(c)(i) Drag $= \frac{1}{2} 1.25v^2 (4\pi r^2)C_D$ ✓ BOD

$0.06 \times mg = = \frac{1}{2} 1.25v^2 (4\pi \times 0.02625^2)C_D$ ✓✓

$v = 12.3 \, m \, s^{-1}$ ✓ BOD (incorrect rounding)

MARKER NOTE
Ffion's answer, once again, is succinct but the answer is correct and all the steps are shown (not that they are required when the answer is correct). She gains full marks once more.

4 marks

(ii) If the forces are equal at such a low speed ✓ (ecf) then it seems to me that air resistance is more important than friction and Archie might just be correct. ✓ (ecf)

MARKER NOTE
Ffion has made an excellent effort at this part too and obtains 2 marks. She compares the two forces (by saying they are equal at a low speed) and then comes to a very sensible conclusion (applying ecf for his low value in the previous part). She cannot gain the 2nd mark because there is nothing here relating to how the drag varies with speed.

2 marks

Total **12 marks /13**

Option D: Energy and the environment

Topic summary

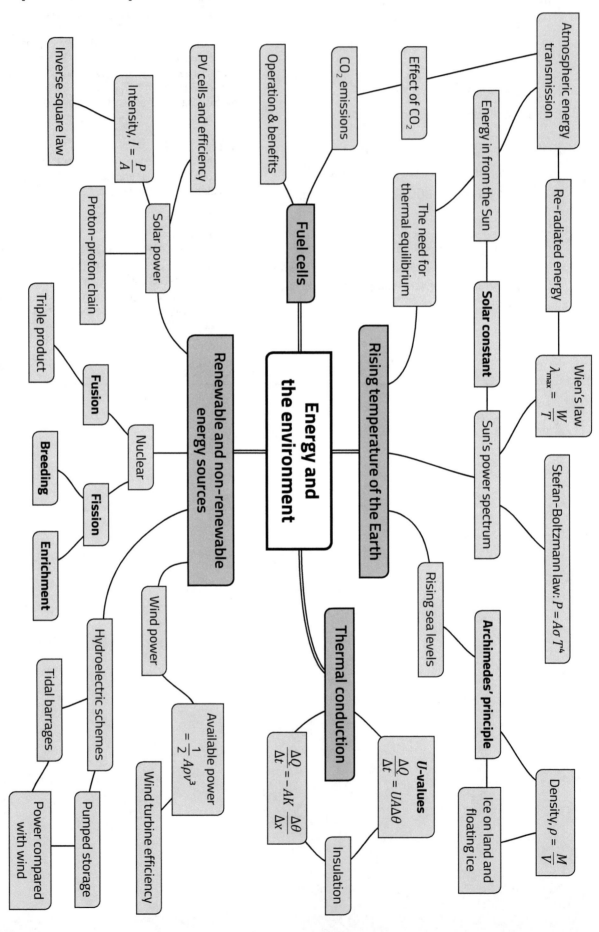

Q1 Asteroid 2010 TK7 has a diameter of 300 m. It orbits the Sun at the same distance as the Earth, 1.50×10^{11} m. The asteroid has no kind of atmosphere.
[Solar constant = 1361 W m^{-2}]

(a) The asteroid reflects 10% of the incident radiation back into space.

   (i)   Calculate the rate of absorption of solar energy by the asteroid.   [2]

   (ii) Assume that the asteroid is rotating and radiates as a black body. Show that the mean surface temperature of 2010 TK7 is approximately 270 K, explaining the significance of the assumptions.   [4]

   (iii) Calculate the peak wavelength of the radiation emitted from the surface of the asteroid, 2010 TK7. Identify the region of the electromagnetic spectrum in which this wavelength lies.   [3]

(b) In spite of reflecting a bigger fraction of the incoming solar radiation, the mean surface temperature of the Earth is higher: approximately 288 K.

   (i)   Explain how the presence of greenhouse gases, such as carbon dioxide, in the atmosphere accounts for this. No calculations are needed.   [4]

   (ii) Explain briefly why increased concentrations of greenhouse gases are causing global temperatures to rise.   [2]

Q2 (a) State the principle of Archimedes. [2]

....

....

....

(b) A block of ice has a volume of 100 cm³. It is lowered gently into a displacement can which is full of sea water.

ρ_{ice} = 920 kg m⁻³; $\rho_{sea\ water}$ = 1028 kg m⁻³

(i) Calculate the volume of seawater which is displaced by the ice. [3]

....

....

....

....

(ii) The ice in the can melts. Explain what happens to the water level in the can and the relevance of this for rising sea levels. [3]

....

....

....

....

(c) The area of ocean covered by sea ice and by glaciers has decreased because of global warming. This loss of ice accelerates global warming. Explain why this is the case. [2]

....

....

....

Q3 A common argument against the reliance on renewable sources for electrical energy generation is that they cannot be relied upon because they are intermittent. Discuss to what extent this is true and suggest how this drawback might be overcome. [4]

....

....

....

....

....

....

Q4 (a) State what is meant by a *fuel cell*. [2]

....

....

....

(b) Discuss the advantages in powering vehicles by fuel cells rather than petrol or diesel engines. [3]

....

....

....

....

....

Q5 (a) State what is meant by uranium *enrichment* and explain why it is necessary. [3]

....

....

....

....

....

(b) Fissile nuclides can be bred (i.e. produced) when a non-fissile nuclide absorbs a neutron from a fission reactor. This produces a radioactive nuclide which undergoes multiple β^- decays to give the desired fissile nuclide.

The fissile $^{233}_{92}U$ can be bred from an isotope of thorium, Th, for which Z = 90.

(i) Write the reactions for breeding $^{233}_{92}U$. [2]

(ii) Describe how a fissile nuclide of plutonium is produced. [3]

....

....

....

....

....

Q6 (a) Air, of density, ρ, travels at a speed, v, along a pipe of cross-sectional area A.
Show that the kinetic energy of the air which passes any point in the pipe per second is given by $\frac{1}{2}A\rho v^3$. A space is included for a diagram, if needed. [3]

(b) A wind turbine with blade length 80 m operates in a steady wind of speed 12 m s⁻¹. It generates 7.9 MW of electrical power of. Calculate its efficiency. [ρ_{air} = 1.25 kg m⁻³] [3]

Q7 The thermal conduction equation can be written:

$$\frac{\Delta Q}{\Delta t} = -AK\frac{\Delta \theta}{\Delta x}.$$

(a) Explain what the following components of the equation refer to and give their units: [3]

$\frac{\Delta Q}{\Delta t}$ ....

....

$\frac{\Delta \theta}{\Delta x}$ ....

....

(b) Hence, show that the unit of thermal conductivity, K, is W m⁻¹ K⁻¹. [2]

(c) State the significance of the minus sign (−) in the equation. [1]

(d) Calculate the heat transfer per minute through a 0.50 m² panel of pine of thickness 12.0 mm if there is a temperature difference of 25 °C across the panel. [3]
[K_{pine} = 0.14 W m⁻¹ K⁻¹]

Q8   A domestic photovoltaic (PV) system consists of a set of individual PV panels. The manufacturer produces the following characteristic graphs for the 2.0 m² individual panels illuminated at 90° by solar radiation at different intensities.

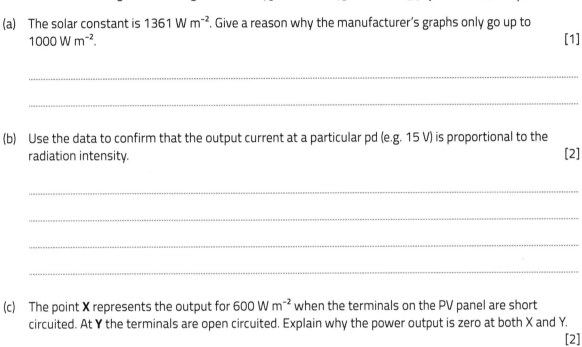

(a)   The solar constant is 1361 W m⁻². Give a reason why the manufacturer's graphs only go up to 1000 W m⁻². [1]

....

....

(b)   Use the data to confirm that the output current at a particular pd (e.g. 15 V) is proportional to the radiation intensity. [2]

....

....

....

....

(c)   The point **X** represents the output for 600 W m⁻² when the terminals on the PV panel are short circuited. At **Y** the terminals are open circuited. Explain why the power output is zero at both X and Y. [2]

....

....

....

(d)   The manufacturer claims that the panel can give out at least 220 W when in light of intensity 600 W m⁻². Evaluate this claim and determine the output pd and current which deliver the maximum power. [4]

....

....

....

....

....

....

Q9 In the small-scale hydroelectric scheme shown, 2.75 m³ of water flows through the turbine each second; the diameter of the outflow pipe is 1.0 m².

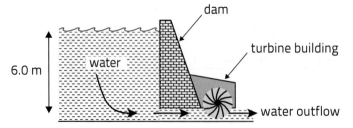

(a) Show that:

(i) the rate of gain of kinetic energy by the water is approximately 17 kW, [3]

(ii) the rate of loss of potential energy by the water is approximately 160 kW. [2]

(b) Assuming that the turbine/generator combination converts 80% of the available energy to electrical energy, calculate the output power of the generator. [3]

(c) Colin claims that if the turbine is adjusted to allow a 10% greater flow rate of water, the power output would be 10% more but overall efficiency of the hydroelectric generation would be 10% less. Evaluate these claims. [5]

Q10 The first two stages of the proton-proton chain are:

$$^1_1H + {^1_1}H \rightarrow {^2_1}H + {^0_1}e + {^0_0}\nu_e \quad \text{and} \quad {^2_1}H + {^1_1}H \rightarrow {^3_2}He + \gamma$$

The mean lifetime of protons in the core of the Sun is over 1000 million years, whereas a deuteron, ^2_1H, reacts to give ^3_2He within about 1 second. Briefly explain this difference by considering the interactions involved in the two stages. [2]

....

....

....

Q11 The exterior of a flat consists of an insulated wall with a U-value of 0.18 W m^{-2} K^{-1} and two identical double-glazed windows with U values of 1.5 W m^{-2} K^{-1}.

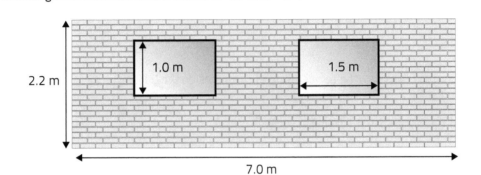

(a) Explain what is meant by a U-value of 0.18 W m^{-2} K^{-1}. [2]

....

....

....

(b) The flat owner is considering replacing the double-glazed units with triple glazing which has a U-value of 0.8 W m^{-2} K^{-1}. Assuming that the flat only loses heat through the wall and windows, calculate the percentage reduction in heating bill which the owner should expect. [3]

....

....

....

....

(c) Suggest why triple-glazed windows are much more common in Norway than in Wales and England. [2]

....

....

....

Q12 The wall of a building consists of two 10 cm brick layers with a 10 cm layer of mineral wool insulation between. On one day, the temperature difference across the interior brick layer is 0.35°C.
$K_{brick} = 0.62$ W m^{-1} K^{-1}; $K_{insulation} = 0.039$ W m^{-1} K^{-1}

(a) Calculate the rate of heat loss from an 8.0 m^2 section of wall. [2]

(b) Show that the temperature difference across the whole wall is approximately 6 °C. [2]

(c) On a windless day, the temperature of the air inside the building is 22 °C and that of the outside air is 8 °C.

(i) Determine the U-value of the wall. [3]

(ii) With the aid of a diagram but without any calculations, explain why the temperature difference between the inside and outside air temperatures is much greater than the 6 °C given in part (b). [3]

(iii) Explain briefly why you would expect the rate of heat loss on a windy day to be greater than predicted by the U-value. [2]

Q13 The graph shows the variation of intensity with wavelength of the Sun's radiation.
[Note the logarithmic scale.]

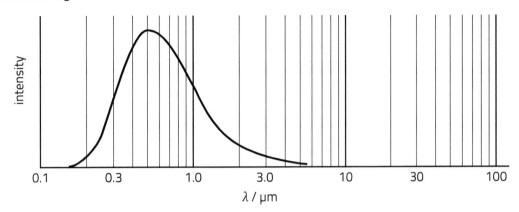

(a) Show that the temperature of the surface of the Sun is approximately 6000 K. [2]

....

....

....

(b) The mean temperature of the Earth's surface is about 290 K. Sketch the emission spectrum of the Earth on the same axes. [The height of the peak is not important.] [2]

[Space for calculations]

(c) Radiation between 0.4 and 0.9 µm passes through the atmosphere with little absorption. Between 3 and 10 µm, a significant fraction of radiation is absorbed by carbon dioxide, water vapour and methane molecules. At longer wavelengths virtually all is absorbed.

Explain the significance of this information for the temperature of the surface of the Earth. [5]

....

....

....

....

....

....

....

....

....

Q14 It is planned to place a thermal store in a cupboard which has a mean temperature of 20°C. The store is to be used for providing hot water for domestic use. The thermal store consists of a stainless steel cylinder of diameter 0.60 m and thickness 0.3 cm, containing 400 litres of water. The cylinder is insulated by a 2.5 cm thick polyurethane (PU) foam jacket and stands on an insulating base.

A domestic PV system is used to maintain the temperature of the water at 65°C, via an immersion heater.

$K_{PU} = 0.025$ W m^{-1} K^{-1}; $K_{steel} = 45$ W m^{-1} K^{-1}; 1 litre = 10^{-3} m^3

(a) Calculate the surface area of the PU foam ignoring the base. [3]

....

....

....

....

....

(b) In calculating the rate of heat loss from the thermal store, Damian commented that you could ignore the effect of the steel. Justify this comment. [2]

....

....

....

(c) (i) Use this information to show the water will need to be heated at a rate of about 130 W by the PV system to maintain its temperature. [2]

....

....

....

(ii) Estimate the temperature difference across the stainless steel and consider again whether ignoring the stainless steel in part (i) was justified. [2]

....

....

....

(d) When installed, it is noted that the actual heating rate when heat is not being extracted from the store for domestic use is only 70 W. Discuss what was ignored in the calculation in (c)(i). [2]

....

....

....

Question and mock answer analysis

Q&A 1

(a) The first stage in the fusion reactions taking place in the Sun is:

$$^1_1H + {}^1_1H \rightarrow {}^2_1H + {}^0_1e + {}^0_0v_e$$

with the release of 2 MeV.

Suggest **two** reasons why this reaction is not considered appropriate for use in a practical nuclear fusion reactor. [2]

(b) An important concept in nuclear fusion is that of *confinement time*, τ_E, defined by:

$$\tau_E = \frac{W}{P_{loss}}$$

where W is the energy density (i.e. the energy per unit volume) of the plasma and P_{loss} is the rate of energy loss per unit volume.

Show that, with this definition, τ_E has the unit of time. [2]

(c) In order to produce useful power, the *triple product* must reach a critical value.

(i) The unit of the triple product can be written $K\,s\,m^{-3}$. Identify the three quantities in the triple product and hence justify this unit. [3]

(ii) The critical value of the triple product depends on the temperature. A school physics book gives the following graph of the variation of the critical value of the triple product with temperature for a deuterium–tritium reactor.

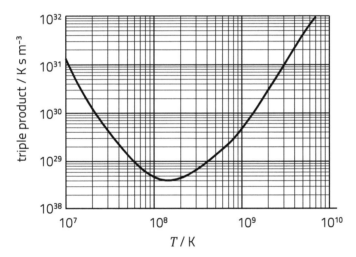

(I) Give the minimum value of the triple product needed to sustain a fusion reaction and the temperature at which this occurs. [2]

(II) In an experimental reactor, a confinement time of 2.0 s is achieved when working at the temperature in part (I). Calculate the density of deuterium and tritium ions necessary to achieve sustained fusion. [2]

Options A–D Practice questions

What the question is asking

There is very little that examiners can ask about nuclear fusion in this option but part (a) is synoptic in nature and is AO3. You are expected to realise that, as this a weak reaction, it is very unlikely to occur and it also gives out very little energy. The triple product is explicitly mentioned in the option specification. Parts (b) and (c) of the question combine knowledge of the triple product with that of SI units and data analysis. Part (b) gives a definition of confinement time and requires its use in deriving a unit – a standard AO2 question. Part (c) (i) extends this to asking for a recall of the definition of triple product (AO1) and justifying its unit (AO2). The data analysis on the triple product in (c)(ii) is quite tricky not only because it introduces the idea that the triple product (which includes temperature) varies with temperature but also because the data is given using a graph with logarithmic scales, which is one of the mathematical requirements.

Mark scheme

Question part			Description	AOs			Total	Skills	
				1	2	3		M	P
(a)			It is a weak interaction so unlikely / fewer interactions (per second)[1] Its energy release is low (cf. ^{2}H + ^{3}H)[1]			2	2		
(b)			$[W] = $ J m^{-3} [or kg m^{-1} s^{-2}] **and** $[P_{loss}] = $ W m^{-3} [or kg m^{-1} s^{-3}] [1] Division seen → s [1] Note: omission of m^{-3} → 1 mark max.		2		2	1	
(c)	(i)		Containment time, particle/ion density, temperature [1] $nT\tau_E$ seen [or in words] [1] m^{-3}, K seen [1]	1 1		1	3	1	
	(ii)	I	4×10^{28} [K s m^{-3}] 1.3×10^8 [K] [allow 1.2 − 1.4]	2			2		
		II	4×10^{28} (ecf) $= n \times 1.3 \times 10^8$ (ecf) $\times$ 2.0 [1] 1.5×10^{20} [m^{-3}]	1	1		2	1	
Total				5	4	2	11	3	0

Rhodri's answers

(a) The protons would just bounce off each other and not react because they are positively charged X [not enough]
2 MeV = 3.2×10^{-13} J X (?)

MARKER NOTE
Rhodri needed to link this to the weak interaction to gain the 1st mark. ^{2}H and ^{3}H ions are also positively charged! What is the significance of 3.2×10^{-13} J?

0 marks

(b) Energy (W) has unit J
Power (P) has unit W X
$W = $ J s^{-1} ∴ $\dfrac{J}{J s^{-1}} = s$ ✓ ecf

MARKER NOTE
Rhodri has not noticed that W and P_{loss} are per unit volume, so loses the first mark. He is allowed the second mark ecf.

1 mark

(c) (i) Apart from the containment time, there is the number of nuclei per cubic metre and the temperature. ✓
The units are s, m^{-3} and K ✓
So we get K s m^{-3} X [not enough]

MARKER NOTE
Rhodri has identified the three variables for the first mark. He has given the units in the same order as the variables so the examiner assumes that the m^{-3} and K are correctly ascribed. He needs to say that the quantities are multiplied together for the middle mark.

2 marks

(ii) (I) Triple product = 3×10^{28} X
Temperature = 0.5×10^8 X

MARKER NOTE
Rhodri has not interpreted the scale correctly, eg the line after 10^8 is 2×10^8 so the temperature at minimum is between 1 and 2×10^8.

0 marks

(II) Density = $\dfrac{3 \times 10^{28}}{0.5 \times 10^8 \times 2.0}$ ✓ (ecf)

$= 3 \times 10^{20}$ m^{-3} ✓

So 1.5×10^{20} each of deuterium and tritium.

MARKER NOTE
The examiner has allowed ecf on the answer to part (b)(ii)I. Dividing the density into deuterium and tritium wasn't required but certainly isn't penalised!

2 marks

Total   **5 marks /11**

Ffion's answers

(a) The neutrino shows that this is a weak interaction. So, if when lots of protons collide very few will react like this and change into ^3_1H so very little energy will be produced. ✓

Also a lot of the energy will be taken away by the neutrinos — this will be lost to the reactor. ✓

MARKER NOTE
This is an excellently developed answer on the first marking point only. Unfortunately the mark scheme does not allow two marks to be given for this. However, Ffion's second point which is not on the mark scheme is very insightful and would have been awarded after consultation with a senior examiner.
2 marks

(b) Energy per unit volume $= kg\ m^2\ s^{-2} \times m^{-3}$
$$= kg\ m^{-1}\ s^{-2}$$
So power loss per unit volume $= kg\ m^{-1}\ s^{-3}$ ✓
So $\left[\dfrac{W}{P_{loss}}\right] = \dfrac{kg\ m^{-1}\ s^{-2}}{kg\ m^{-1}\ s^{-3}}$ ✓ $= s$

MARKER NOTE
The answer is fine, though Ffion didn't need to express [W] and [P_{loss}] in base SI units.
2 marks

(c)(i) The triple product:
Containment time, $\tau_E - s$
number density, $n - m^{-3}$ [bod]
temperature, $T - K$ ✓✓
So $[T\tau_E n]$ ✓ $= K\ s\ m^{-3}$

MARKER NOTE
The only weakness in Ffion's answer is that she doesn't say what the 'number' is. Either reacting nuclei or ions is expected — hence the bod.
3 marks

(ii)(I) Minimum value $= 4 \times 10^{28}\ K\ s\ m^{-3}$ ✓
Temperature $= 1.5 \times 10^8\ K$ ✗

MARKER NOTE
Ffion has basically understood the way that a logarithmic scale works and got the easy first mark. She has not followed through with applying the non-linearity of the scale in the more difficult second mark.
1 mark

(II) $4 \times 10^{28} = 1.5 \times 10^8 \times 2.0\ n$ ✓ (ecf)
$\therefore n = 1.3 \times 10^{20}\ m^{-3}$ ✓

MARKER NOTE
The examiner has allowed ecf on the incorrect temperature, so full marks.
2 marks

Total   **10 marks /11**

Options A–D Practice questions

Practice papers

A LEVEL PHYSICS
UNIT 3 PRACTICE PAPER
[Section A only]

1 hour 35 minutes

For Examiner's use only		
Question	Maximum Mark	Mark Awarded
1.	11	
2.	18	
3.	12	
4.	13	
5.	12	
6.	14	
Total	80	

Notes

A WJEC Unit 3 paper has two sections and a total time of 2 hours. The questions on the following pages represent a practice version of Section A. Section B consists of a passage of information on a physics topic followed by questions totalling 20 marks. You are recommended to practise for Section B by using past WJEC and Eduqas papers from the current and previous specifications.

The following information will be given on the front of a WJEC paper:

1. **Additional materials**
   You will be told that you will require a calculator and a Data Booklet. Sometimes you will be told that you need a ruler and/or an angle measurer / protractor.

2. **Answering the examination**
   You will be told to use a blue or black ball-point (but graphs are best drawn using a pencil).
   You will be told to answer all the questions in the spaces provided on the question paper.

3. **Further information**
   Each question part shows, using square brackets, the total marks available. One question will assess the quality of extended response [QER]. This question will be identified on the front page. In this practice paper the QER question is question **3(a)**.

SECTION A

*Answer **all** questions.*

1   (a)   (i)   Define *acceleration*.   [1]

          (ii)   Give an example of a body that is not accelerating.   [1]

    (b)   A car is being driven round a level circular track of diameter 120 m at a constant speed. Each lap takes 24.0 s. Calculate:

          (i)   the car's angular velocity about the centre of the track;   [2]

          (ii)   the car's speed;   [2]

          (iii)   the car's acceleration.   [2]

    (c)   The diagram shows the car at one point in its travel:

view from above (not to scale)

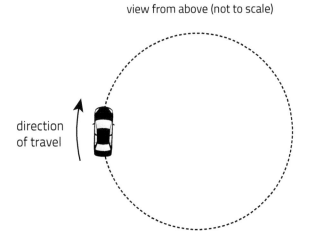

direction
of travel

Put arrows on the car to show the directions of:

- The resultant force on the car. Label the arrow 'R'.   [1]
- The air resistance (drag) on the car. Label the arrow 'D'.   [1]
- The horizontal force on the car from the road. Show as a single arrow labelled 'F'.   [1]

2. (a) A sphere of mass 0.20 kg hangs from a helical spring whose top is clamped. The sphere is pulled down below its equilibrium position and released at time $t = 0$, so that it oscillates up and down. A displacement–time graph is given:

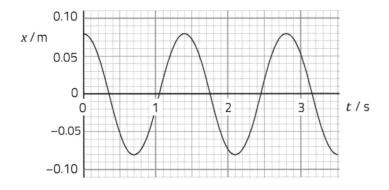

(i) Determine the stiffness constant, k, of the spring. [3]

....

....

....

....

....

(ii) Calculate the maximum kinetic energy of the sphere. [3]

....

....

....

....

(iii) Glyn was asked to draw a sketch-graph of kinetic energy, E_k, against time, t, for the oscillating mass. A vertical scale was not required. Evaluate his attempt (below). [3]

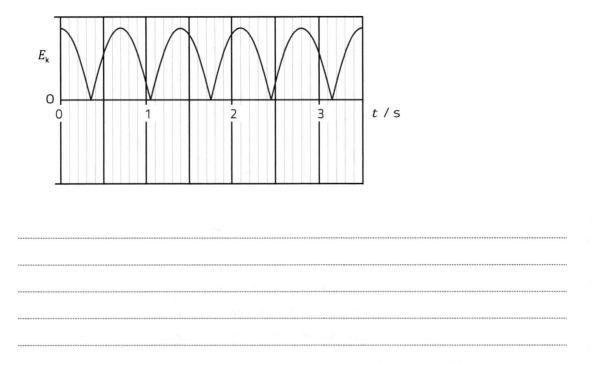

....

....

....

....

....

(b) A light piece of paper is now attached to the sphere, and the apparatus of part (a) is now used to investigate damped oscillations. The sphere is pulled down 60 mm below its equilibrium position and released at time $t = 0$. The amplitude, A, of the oscillations is measured at intervals of 2 cycles, and points are plotted on a grid of A against t:

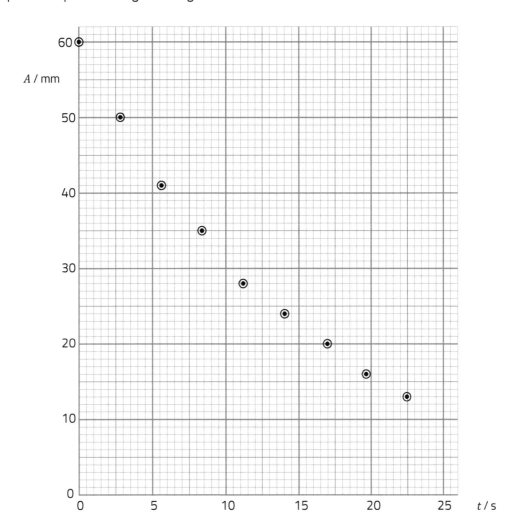

(i) Describe how, after a given number of oscillations, you could determine A using simple laboratory equipment. Include precautions that you would take to obtain a result that is accurate. [3]

....

....

....

....

....

(ii) The amplitude is expected to decrease with time according to the equation:

$$A = A_0 e^{-\left(\frac{t}{\tau}\right)}$$

in which A_0 and τ are constants.
Make use of the plotted points to obtain a value for τ. [4]

....

....

....

....

....

(iii) Explain briefly how you could use the measured pairs of values of A and τ to obtain a more repeatable value for τ. [2]

....

....

....

3   (a)   A partially inflated balloon slowly expands when taken into a warmer room. Give a step-by-step
explanation of why this happens *in terms of molecules and the kinetic theory*.   [6 QER]

(b)   A cylinder of volume 0.025 m³ contains 6.0 mol of helium, a monatomic gas, at a pressure of 600 kPa.
(M_r of helium = 4.0)

(i)   Calculate the rms speed of the molecules.   [3]

(ii)   The cylinder is now moved to a store in which the temperature is 285 K. Determine whether the
rms speed of the helium molecules will increase, decrease or stay the same.   [3]

4. (a) The first law of thermodynamics may be written as:

$$\Delta U = Q - W$$

Explain in terms of energy what Q represents in this equation. [2]

....

....

....

(b) The diagram shows a change, AB, undergone by a fixed amount of ideal gas, initially at a temperature of 280 K:

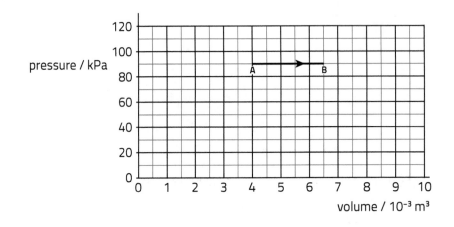

(i) Determine the amount of gas in mol. [2]

....

....

....

(ii) Show that the temperature increase between A and B is approximately 180 K. [2]

....

....

....

(iii) Use the first law of thermodynamics to determine the heat flow during this change. [5]

....

....

....

....

....

....

(iv) The diagram shows another possible route by which the gas can be taken from A to B:

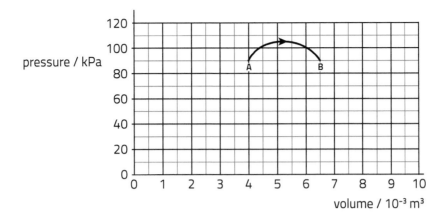

Explain whether the heat flow will be greater, the same or less than that calculated in (b) (iii).   [2]

5. (a) Define the *activity*, A, of a sample of a radioactive isotope. [1]

    ....

    ....

    ....

    (b) Use the radioactive decay equation, $A = A_0 e^{-\lambda t}$, to show that the decay constant, λ, is related to the half-life, $T_{1/2}$, by the equation:

    $$\lambda = \frac{\ln 2}{T_{1/2}}$$ [2]

    ....

    ....

    ....

    (c) $^{32}_{15}P$ is a β^- emitter, with a half-life of 14.2 days, that decays to an isotope of sulfur (S).

    (i) Complete the nuclear equation for this decay, including all decay products: [2]

    $$^{32}_{15}P \rightarrow$$

    (ii) A sample of $^{32}_{15}P$ has activity 240 GBq.

    (I) Calculate the mass of the sample, giving your answer in an SI unit. [4]

    ....

    ....

    ....

    ....

    ....

    ....

    (II) Rhian claims that the activity of the sample will decrease to 180 Bq in less than 7.1 days. Evaluate her claim. [3]

    ....

    ....

    ....

    ....

    ....

    ....

6.   For this question you will need these data:

mass of proton   = 1.00728 u

mass of neutron   = 1.00866 u

mass of $^7_3$Li nucleus = 7.01435 u

mass of $^4_2$He nucleus = 4.00151 u

1 u ≡ 931 MeV.

(a) (i)   Calculate the binding energy **per nucleon** of a $^7_3$Li nucleus.   [3]

....

....

....

....

....

(ii) (I)   Sketch a graph of binding energy per nucleon against nucleon number on the axes provided. Scales are not needed.   [1]

(II) Use the graph to explain why energy is released when light nuclei fuse or heavy nuclei undergo fission.   [3]

....

....

....

....

....

(b) A proton with a kinetic energy of 0.800 MeV collides with a stationary $^{7}_{3}$Li nucleus, and causes this disintegration:

$$^{1}_{1}p + ^{7}_{3}Li \longrightarrow ^{4}_{2}He + ^{4}_{2}He$$

(i) State the pd needed to accelerate the proton from rest to give it a kinetic energy of 0.800 MeV. [1]

....

....

....

(ii) Calculate the sum of the kinetic energies of the $^{4}_{2}$He nuclei. [3]

....

....

....

....

(iii) One of the $^{4}_{2}$He nuclei is emitted in the same direction in which the proton was travelling. Explain why the $^{4}_{2}$He nuclei cannot share the kinetic energy equally. [3]

....

....

....

....

....

A LEVEL PHYSICS
UNIT 4 PRACTICE PAPER

2 hours

		For Examiner's use only	
	Question	Maximum Mark	Mark Awarded
Section A	1.	16	
	2.	15	
	3.	11	
	4.	9	
	5.	11	
	6.	10	
	7.	8	
Section B	Option	20	
	Total	100	

Notes

The WJEC Unit 4 examination paper has two sections, A and B. Section A consists of core questions totalling 80 marks – you are advised to spend about 1 hour 35 minutes on these. Section B consists of questions on four options, each out of 20 marks: you should answer only one of these option questions and are advised to spend about 25 minutes on this question.

The following information will be given on the front of a WJEC paper:

1. **Additional materials**
   You will be told that you will require a calculator and a **Data Booklet**. Sometimes you will be told that you need a ruler and/or an angle measurer / protractor.

2. **Answering the examination**
   You will be told to use a blue or black ball-point (but graphs are best drawn using a pencil).
   You will be told to answer all the questions in the spaces provided on the question paper.

3. **Further information**
   Each question part shows, using square brackets, the total marks available. One question will assess the quality of extended response [QER]. This question will be identified on the front page. In this practice paper the QER question is question **5(b)**.

SECTION A

*Answer **all** questions.*

1. (a) Sketch diagrams of the two ways in which a 2.0 μF capacitor and a 3.0 μF capacitor can be connected
together, and determine the capacitance of each combination. [4]

Combination (1)   Combination (2)

....

....

....

....

....

(b) Fred is designing an electric toothbrush that will run off a charged capacitor. The capacitor is to be
charged to a pd of 6.0 V and connected across the toothbrush motor, which will work effectively until
the pd has dropped to 4.5 V. The mean power taken by the motor is 0.75 W. Fred believes that a 10 F
capacitor, charged to 6.0 V, will supply the energy needed for 2.0 minutes of tooth cleaning. Evaluate
his belief. [3]

....

....

....

....

....

(c) Jules investigates the discharging of a capacitor through a resistor, using the circuit shown, which
includes a 3-position switch, S.

He charges the capacitor (switch to the left), allows the capacitor to discharge (switch to the right) for a measured time, t, and reads the pd, V, (switch central). He repeats the procedure so as to be able to plot points with error bars for a graph of ln (V/V) against t (see below).

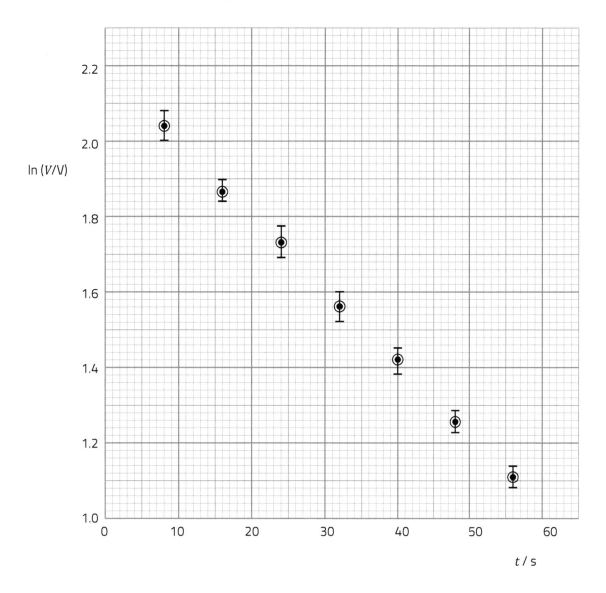

(i) Starting with the equation $Q = Q_0 e^{-t/CR}$, show that the expected relationship is:

$$\ln(V/V) = \ln(V_0/V) - \frac{1}{CR} t$$

[2]

(ii) For t = 24 s, the following readings of pd were taken:

V / V: 5.43, 5.52, 5.74, 5.88

Evaluate whether or not the error bar has been drawn correctly for t = 24 s. [2]

(iii) Make use of the plotted data to determine the capacitance of the capacitor, together with its absolute uncertainty. [5]

2. (a) Define the *electric field strength, E,* at a point. [1]

....

....

....

(b) The electric field strength at a distance of 0.10 m from a small, positively charged, sphere is 2.0 N C^{-1}.
Show that the charge, Q, on the sphere is approximately 2 pC. [2]

....

....

....

(c) The sphere in part (b) and an identical sphere with the same charge are placed as shown:

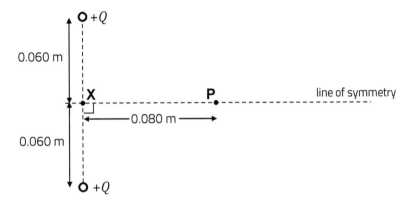

Determine the magnitude and direction of the electric field strength at point P due to the charges.
You may add to the diagram. [4]

....

....

....

....

....

....

(d) An ion of mass 1.05×10^{-25} kg and charge 1.60×10^{-19} C, initially at rest at **X**, is given a very small
initial velocity to the right. Determine the maximum speed that it reaches. [4]

....

....

....

....

....

....

(e) Elodie gives this description of the ion's motion. 'As the ion leaves X and travels further and further to the right its acceleration gets less and less.' Evaluate to what extent Elodie's description is accurate. Calculations are not wanted. [4]

....

....

....

....

....

....

....

3. (a) A graph of radial *velocity* against time is given for the star 18 Delphini. The *mean* radial velocity has been subtracted, leaving just the star's changing velocity due to an orbiting planet.

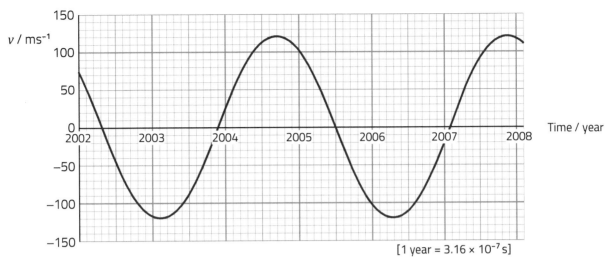

[1 year = 3.16 × 10⁻⁷ s]

(i) Determine the radius of the **star's** orbit. [3]

....

....

....

....

....

(ii) The star's mass has been determined to be 4.67×10^{30} kg. Show that the star and planet are roughly 4×10^{11} m apart, stating one assumption or approximation that you are making. [4]

....

....

....

....

....

....

(iii) Determine the mass of the planet. [2]

....

....

....

....

Unit 4 and Options Practice Paper

(b) Astronomers have discovered planets orbiting stars other than the Sun that may have similar surface conditions to the Earth. It has been suggested that funding for searching out such 'Earth-like exoplanets' should be especially generous. Discuss this suggestion. [2]

....

....

....

....

....

4. (a) State what is meant by the *Hubble constant, H_0*. [1]

....

....

    (b) Use the Hubble constant to answer the following:

        (i) A line in the hydrogen atomic spectrum has a wavelength of 656 nm as measured in the laboratory. The same line is observed in light reaching us from a distant galaxy, but with a wavelength of 694 nm. Determine the distance of the galaxy from us. [3]

....

....

....

....

....

        (ii) Estimate the age of the universe and give one reason (apart from uncertainty in the value of H_0) why this value is only an estimate. [2]

....

....

....

    (c) The *critical density, ρ_c*, of a 'flat' universe if given by:

$$\rho_c = \frac{3H_0^2}{8\pi G}$$

Show that this equation is correct as far as units (or dimensions) are concerned. [3]

....

....

....

....

....

....

5. (a) Two students, Chioke and Emeka, have been given the task of determining the magnetic field strength, B, in a region, using the apparatus shown, and a set of small standard masses.

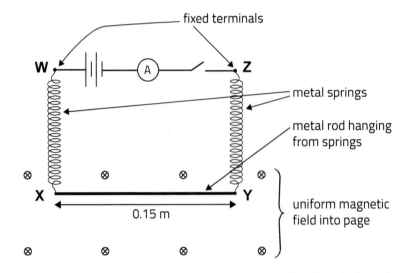

(i) The students require the metal rod, XY, to move upwards when the switch is closed. Evaluate whether, for this to happen, the battery is the right way round, or whether it needs to be reversed. [2]

....

....

....

(ii) With the battery the right way round and the switch closed, the ammeter reads 4.20 A. The students find that a mass of 4.50 gram, hung from the middle of the rod, will bring it back down to its position before the switch was closed. Determine the magnetic field strength. [3]

....

....

....

....

....

(b) The students are invited to investigate electromagnetic damping using a modified version of the apparatus in (a).

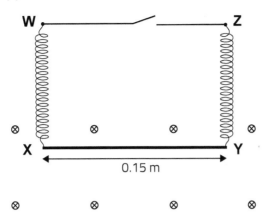

First with the switch open, then with the switch closed, they displace the rod downwards by a small distance and observe the rod's oscillations up and down. They see that the oscillations are more highly damped when the switch is closed, and correctly attribute this to an electromagnetic force on the rod.

Explain carefully how the electromagnetic force arises, and how we account for the direction in which it acts.
[6 QER]

6.   The diagram shows a battery driving current from left to right through the Hall wafer. A uniform magnetic field, B, is applied at right angles to the large faces of the wafer, as shown:

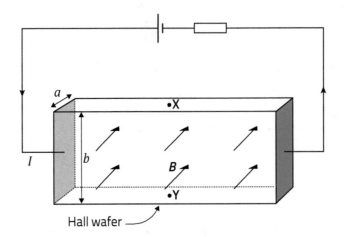

(a) The Hall voltage, V_H, between **X** and **Y** is given by the equation:

$$V_H = bBv$$

In which v is the drift speed of the charge carriers in the wafer.

Explain how the Hall voltage arises, and derive the equation given above. The charge carriers in the wafer are positive.   [5]

....

....

....

....

....

....

....

....

....

(b) For a given current, I, and a given magnetic flux density, B, the Hall voltage, V_H, is greater when a, (the wafer thickness) and n (the number of charge carriers per unit volume) are smaller. Explain why this is so.   [2]

....

....

....

(c) The Hall wafer, carrying a fixed current, is placed a measured distance, r, from a long, straight, current-carrying wire. The wafer is orientated so that the wire's magnetic field is at right angles to its largest faces, and V_H is measured. The procedure is repeated for two more distances, and also with the current in the wire turned off. Here are the results:

r / mm	40	60	80	No current in wire
V_H / mV	79	57	47	15

Examine whether or not these readings support the expected variation with distance of the wire's magnetic field. [3]

....

....

....

....

....

....

7. (a) Draw magnetic field lines on the diagram to show the magnetic field inside and beyond the ends of the solenoid. [3]

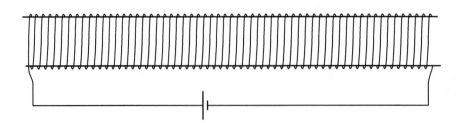

(b) A 400-turn solenoid of length 0.80 m is connected to a signal generator producing a triangular waveform. When the current through the solenoid is rising, it does so at a rate of 0.60 A s⁻¹.

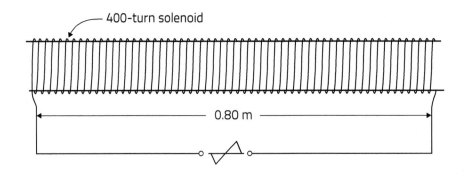

(i) Show that the rate of rise of magnetic flux density in the central region of the solenoid is approximately 4×10^{-4} T s⁻¹. [2]

....

....

....

(ii) A short coil, of smaller diameter than the solenoid, is now placed centrally inside the solenoid so that its axis coincides with the solenoid's axis. The short coil has 250 turns and its **diameter** is 30 mm. Calculate the emf induced in the short coil when the current in the solenoid is rising. [3]

....

....

....

....

....

....

SECTION B: OPTIONAL TOPICS

Option A – **Alternating currents**

Option B – **Medical physics**

Option C – **The physics of sports**

Option D – **Energy and the environment**

Answer the question on one topic only.

Place a tick (✓) in one of the boxes above, to show which topic you are answering.

You are advised to spend about 25 minutes on this section.

Option A – Alternating currents

8. (a) A 2.5 kW electric kettle operates from the 230 V_{rms} 50 Hz mains supply.

On the axes below, sketch a graph of the power, P, transferred to the heating element from the mains over a period of 40 ms. Include appropriate values on the power axis. [3]

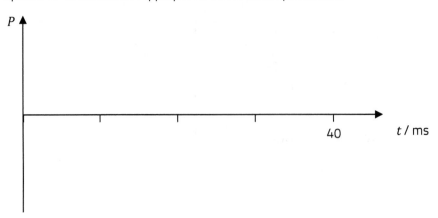

Space for calculations (if needed):

(b) A capacitor, C, has a reactance of 40 Ω at 500 Hz. It is connected in series with a 30 Ω resistor across a 12 V rms, 500 Hz power supply.

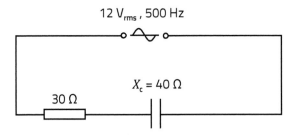

(i) Show that the rms pd across the resistor and capacitor are 7.2 V and 9.6 V respectively. [3]

....

....

....

....

....

....

(ii) Explain briefly how the sum of the voltages in part (i) can be more than the rms voltage of the supply. [1]

....

....

(iii) Jamie claims that if an inductor is added in series to the resistor and capacitor, the pd across the resistor will increase. Evaluate his claim. [3]

....

....

....

....

....

(c) In order to measure the frequency and rms current in a 12 kΩ resistor in different AC circuit, Nigel connected an oscilloscope across the resistor. He initially obtained trace 1. By adjusting the oscilloscope settings, he obtained trace 2.

trace 1	trace 2
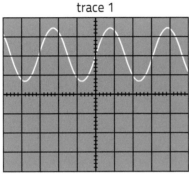	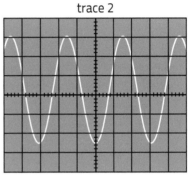

(i) Describe what changes Nigel made to the oscilloscope settings and explain how these changes allowed him to obtain better readings for determining the current. [3]

....

....

....

....

....

(ii) Nigel used the following settings for trace 2:
Time base: 20 μs div^{-1} Y-gain: 10 mV div^{-1}

Determine:

(I) the frequency of the AC; [2]

....

....

....

(II) the rms current in the resistor. [3]

....

....

....

....

....

(iii) Gareth suggested that a time base setting of 10 ms div^{-1} would have been better. Comment briefly on his suggestion. [2]

....

....

....

Option B – Medical physics

9. (a) An ultrasound A-scan is performed on an eye. The diagram shows the arrival times at the piezo-electric transducer of a pulse reflected by some parts of the eye:

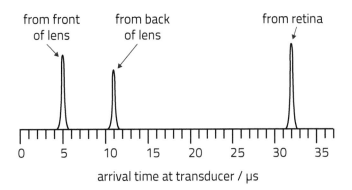

(i)   The back of the lens is separated from the retina by a substance call *vitreous humour* (VH) in which the speed of ultrasound is 1620 m s^{-1}. Calculate the distance between the back of the lens and the retina.   [3]

(ii) Determine the percentage of ultrasound intensity reflected by the boundary between the lens and the VH, using the equation:

$$\frac{I_{reflected}}{I_{incident}} = \frac{(Z_2 - Z_1)^2}{(Z_2 + Z_1)^2}$$

Acoustic impedance of lens = 1.74 units
Acoustic impedance of VH   = 1.53 units   [2]

(iii) To do the scan the ultrasound transducer is held against the front of the eye. Explain why a coupling agent is needed between the transducer and the eye.   [2]

(b) The diagram shows an X-ray spectrum from an X-ray tube with a molybdenum target.

(i) Calculate the accelerating pd that was used to produce this spectrum. [3]

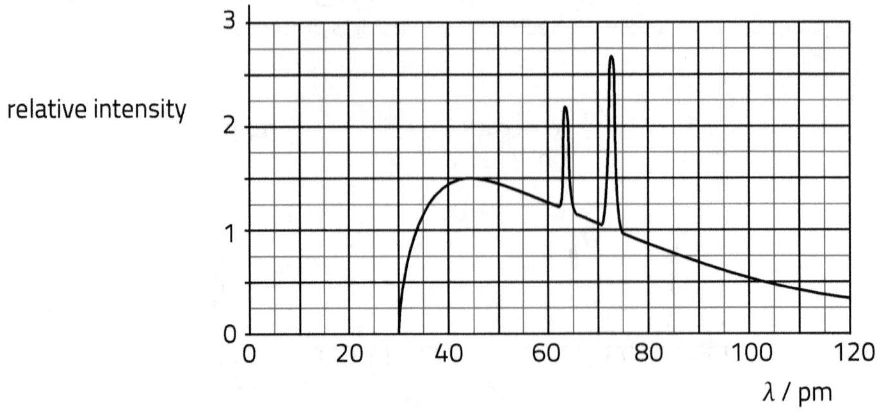
relative intensity
3
2
1
0
0
20
40
60
80
100
120
λ / pm

....

....

....

....

(ii) Explain briefly how the spikes on the spectrum arise. [2]

....

....

....

....

(c) (i) Calculate the Larmor frequency for ${}^{1}_{1}H$ nuclei in a magnetic field of 1.50 T. [2]

....

....

....

....

(ii) Explain why a non-uniform magnetic field is used in an MRI scanner. [2]

....

....

....

....

(iii) Explain how an MRI scanner distinguishes different tissues. [2]

....

....

....

....

(d) Soft tissue can be imaged using a CT scan or using an MRI scan. Give one advantage and one disadvantage of an MRI scan over a CT scan. [2]

....

....

....

Option C – The physics of sports

10. (a) Paddle boards are designed for people to use on the water. Users stand on the top of the floating board.

Paddle boards are rather unstable and beginners often find themselves in the water, rather than on the board. Paddle boards are available in various widths and beginners are advised to start on wide ones.

Explain this advice – you may add to the diagram to help your explanation. [3]

....

....

....

....

....

(b) The maximum grip of car tyres on roads is approximately equal to the normal contact force exerted by the road on the tyres. Formula 1 (F1) cars have additional features called wings (or spoilers) to help with their performance. The diagram shows the airflow around the wings on an F1 car, which is travelling to the right.

rear wing   front wing

(i) Use Newton's laws of motion to explain why there is an increased normal contact force, due to the wings. [5]

....

....

....

....

....

....

....

(ii) F1 racing involves accelerating, decelerating and travelling around corners. Explain why an increased normal contact force improves the performance of the cars. [4]

....

....

....

....

....

....

....

(c) To score a goal in netball the goal shooter has to get the ball through a 38 cm diameter metal hoop which is 3.05 m above the ground. Goal shooters are advised to aim for the apex of an imaginary cone above the hoop.

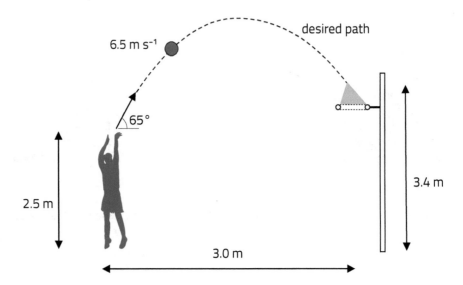

Glenys is hoping that, with her 6.50 m s⁻¹ shot at 65° to the horizontal, the ball will hit the apex of the 3.40 m high cone, and so fall through the hoop. She is standing 3.00 m from the hoop and the launch height is 2.50 m.

(i) Calculate the time it will take the ball to travel 3.00 m horizontally.   [3]

....

....

....

....

....

(ii) Evaluate whether Glenys's shot is likely to be successful. Show your reasoning.   [5]

....

....

....

....

....

....

....

....

Option D – Energy and the environment

11. (a) Tidal stream turbines function just like wind turbines. They can be sited under the sea surface in places where the changing tide causes strong currents.

The power available to a turbine of area A from a tidal stream of speed v is given by:

$$P = \frac{1}{2} A \rho v^3$$

(i) Explain why a tidal stream turbine is significantly smaller than a wind turbine for a similar power output. [2]

....

....

....

....

(ii) A tidal stream generator is proposed at the entrance to a sea loch, which is a tidal inlet connected to the sea by a narrow stretch of water. The proposed site of the generator is shown on the diagram. The rising and falling tides in the sea produce tidal streams into and out of the loch.

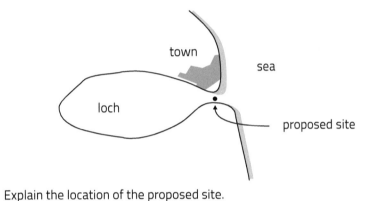

(I) Explain the location of the proposed site. [2]

....

....

....

....

(II) Given that there are suitable sites for a wind farm, explain the advantages of a tidal stream turbine for supplying electricity to the nearby town. [2]

....

....

....

....

(III) Give a reason why, without additional resources, a tidal stream generator could not provide the sole electricity supply to the town. [1]

....

....

(IV) With a tidal stream of 8.0 m s^{-1}, the planned power output of the generator is 3.0 MW. Assuming an efficiency of 35%, calculate the length of the turbine blades required. [Density of sea water = 1020 kg m^{-3}] [2]

....
....
....

(b) (i) Describe briefly the processes of nuclear fission and nuclear fusion. [4]

....
....
....
....
....
....

(ii) Outline the main difficulties in producing sustained nuclear fusion. [3]

....
....
....
....
....

(c) A pharmaceutical company is building a small 'cold store' in which to keep vaccines at a steady –5.0 °C. The design consists of a 2.0 m cube of concrete of thickness 3.5 cm, which is to stand on a well-insulated base in a warehouse which is maintained at 15.0 °C.

The store will include a refrigerating unit which is capable of extracting heat at a rate of up to 1 kW.

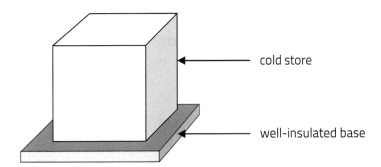

An A level physics student on work experience suggested that putting a 2.5 cm thick layer of polyurethane (PU) foam insulation around the store would allow the refrigeration unit to maintain the inside temperature at –5.0 °C.

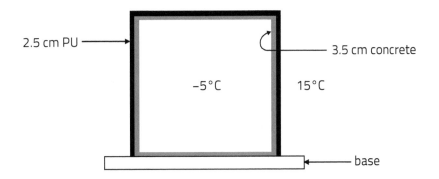

Evaluate the student's suggestion. [4]

$[K_{concrete} = 0.92\ \text{W m}^{-1}\text{°C}^{-1};\ K_{PU} = 0.034\ \text{W m}^{-1}\text{°C}^{-1}]$

....

....

....

....

....

....

....

Answers

Practice questions: Unit 3: Oscillations and Nuclei

Section 3.1: Circular motion

Q1 The angle θ, expressed in radians, is defined by: $\theta = \dfrac{l}{r}$ (see diagram). In this case $\dfrac{l}{r} = 1.2$.

The circumference of a circle $= 2\pi r$, so $360° = 2\pi$ rad,

i.e. $(\theta\,/\,°) = \dfrac{360}{2\pi}\,(\theta\,/\,\text{rad})$

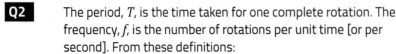

Q2 The period, T, is the time taken for one complete rotation. The frequency, f, is the number of rotations per unit time [or per second]. From these definitions:

If there are N rotations in a time t: $T = \dfrac{t}{N}$ and $f = \dfrac{N}{t}$. Hence $f = \dfrac{1}{T}$.

Q3 (a) [For an object travelling in a circle] the angular velocity is the angle swept out by the radius per unit time.

(b) $\omega = \dfrac{1400 \times 2\pi \text{ rad}}{60 \text{ s}} = 150 \text{ rad s}^{-1}$ (2 sf)

Q4 (a) Tension in rope = centripetal force on the object

$$= \frac{mv^2}{r} = \frac{65 \text{ kg} \times (23.2 \text{ m s}^{-1})^2}{4.5 \text{ m}}$$

$$= 7\,800 \text{ N (2 sf)}$$

(b) The rope does work on the object because the direction of motion (towards the centre) is the same as the direction of the force. Hence energy is transferred to the object. Hence its kinetic energy increases, i.e. it speeds up.

Q5 (a) Acceleration and resultant force are both directed towards the centre of the circle. The force is provided by the [sideways] grip of the road on the car tyres.

(b) $\dfrac{m(v_{max})^2}{r} = mg$, so $v_{max} = \sqrt{rg} = \sqrt{24.0 \times 9.81}$ m s^{-1} = 15.3 m s^{-1} (3 sf)

(c) From part (b) $v_{max} = \sqrt{rg}$, so as r increases so does v_{max} and this claim is correct.

$mr(\omega_{max})^2 = mg$, so $\omega_{max} = \sqrt{\dfrac{g}{r}}$, so as r increases ω_{max} decreases and this claim is also correct.

Q6 (a) (i) $\omega = \dfrac{2\pi \text{ rad}}{(29.5 \times 86\,400 \times 365.25) \text{ s}} = 6.75 \times 10^{-9} \text{ rad s}^{-1}$

(ii) $\omega = \dfrac{2\pi \text{ rad}}{(10.7 \times 3600) \text{ s}} = 1.63 \times 10^{-4} \text{ rad s}^{-1}$

(b) (i) Orbital speed, $v = r\omega = 1.43 \times 10^9 \text{ km} \times 6.75 \times 10^{-9} \text{ rad s}^{-1} = 9.65 \text{ km s}^{-1}$

(ii) Rotational speed, $v = r\omega = 60\,000 \text{ km} \times 1.63 \times 10^{-4} \text{ rad s}^{-1} = 9.78 \text{ km s}^{-1}$

(c) (i) Centripetal acceleration at equator $= r\omega^2 = 6.51 \times 10^{-5} \text{ m s}^{-2}$ [or 6.52 – rounding]

Centripetal force $= ma = 3.70 \times 1022$ N provided by the gravitational force of the Sun

(ii) Centripetal acceleration $= r\omega^2$

$$= 1.59 \text{ m s}^{-2}$$

∴ Percentage reduction in measured $g = \dfrac{1.59 \text{ m s}^{-2}}{10.4 \text{ m s}^{-2}} \times 100\% = 15\%$ (2 sf)

Q7 (a) The centripetal force is provided by the horizontal component of the tension in the pendulum string.

(b) Vertical component of tension, T = weight of the bob

∴ $T\cos 20° = 0.078 \text{ kg} \times 9.81 \text{ N kg}^{-1}$

∴ $T = 0.8143 \text{ N} = 0.81 \text{ N}$ (2 sf)

$T\sin 20° = \dfrac{mv^2}{r}$, so $v = \sqrt{\dfrac{0.153\text{ m} \times 0.8143\text{ N} \times \sin 20°}{0.078\text{ kg}}} = 0.739\text{ m s}^{-1}$

Alternative answer:

Resolve vertically: $T\cos\theta = mg$, so $T = \dfrac{mg}{\cos\theta}$

Horizontally: $T\sin\theta = \dfrac{mv^2}{r}$. Substituting for $T \longrightarrow \dfrac{mg\sin\theta}{\cos\theta} = \dfrac{mv^2}{r}$

∴ Simplifying: $v = \sqrt{rg\tan\theta} = \sqrt{0.153\text{ cm} \times 9.81\text{ m s}^{-2} \times \tan 20°} = 0.739\text{ m s}^{-1}$

Q8 (a) The (inward) gravitational force on the smaller body $= \dfrac{GMm}{r^2}$, where m is the mass of the less massive object. This provides the centripetal force, $\dfrac{mv^2}{r}$

So $\dfrac{mv^2}{r} = \dfrac{GMm}{r^2}$. Multiplying by r and dividing by $m \longrightarrow v^2 = \dfrac{GM}{r}$ QED

(b) Without dark matter $v^2 \propto \dfrac{1}{r}$, so $\left(\dfrac{v_1}{v_2}\right)^2 = \dfrac{r_2}{r_1} = $ constant.

$\left(\dfrac{4700}{3400}\right)^2 = 1.9$. This is very close to 2, so this result is consistent with the absence of dark matter.

This is just a single galaxy, so the data are interesting but not definitive.

Section 3.2: Vibrations

Q1 (a) (i) 0.040 m

(ii) $\omega = \dfrac{2\pi}{T} = \dfrac{2\pi}{1.20\text{ s}} = 5.24\text{ rad s}^{-1}$

(iii) $-\dfrac{\pi}{2}$

(b) $v_{max} = A\omega = 0.040\text{ m} \times 5.24\text{ rad s}^{-1} = 0.21\text{ m s}^{-1}$
Initial velocity is maximum positive, so a cosine graph.

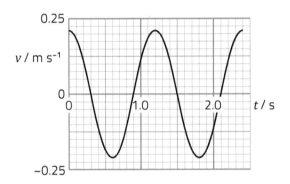

Q2 (a) (i) $x = A\cos\omega t = 0.140 \times \cos\left(\dfrac{2\pi}{0.800\text{ s}} \times 0.50\text{ s}\right)$

$= -0.099\text{ m}$

(ii) $0.070\text{ m} = \frac{1}{2}A$, so $\cos\omega t = 0.5$ and $\omega t = \dfrac{\pi}{3}$

So $\dfrac{2\pi}{0.800\text{ s}}t = \dfrac{\pi}{3}$, ∴ $t = 0.133\text{ s}$

From the graph, the 2nd occasion $= 0.800 - 0.133\text{ s}$
$= 0.667\text{ s}$

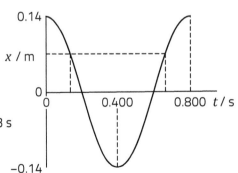

(b) (i) $v = -A\omega\sin\omega t$

$= -0.140 \times \dfrac{2\pi}{0.800\text{ s}} \times \sin\left(\dfrac{2\pi}{0.800\text{ s}} \times 0.50\text{ s}\right)$

$= 0.778\text{ m s}^{-1}$

(ii) $v_{max} = A\omega = 0.140 \times \dfrac{2\pi}{0.800\text{ s}} = 1.10\text{ m s}^{-1}$.

∴ $\sin\omega t = 0.5$, ∴ $\dfrac{2\pi}{0.800}t = \dfrac{\pi}{6}$, ∴ $t = 0.067\text{ s}$

From the graph, the 2nd occasion $= 0.400 - 0.067\text{ s}$
$= 0.333\text{ s}$

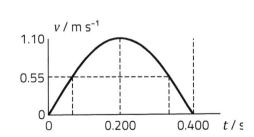

Q3 (a) The (stiffness of the) ruler.

(b) $\omega = 2\pi f = \sqrt{\dfrac{k}{m}}$, $\therefore k = 4\pi^2 f^2 m = 4\pi^2 \times (0.40 \text{ Hz})^2 \times 0.20 \text{ kg} = 1.3 \text{ N m}^{-1}$ [or kg s^{-2}]

(c) $v_{max} = A\omega = 2\pi A f$

$= 2\pi \times 0.050 \text{ m} \times 0.40 \text{ Hz}$

$= 0.13 \text{ m s}^{-1}$

$T = \dfrac{1}{0.4 \text{ Hz}} = 2.5 \text{ s}$

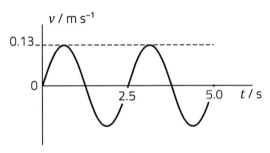

Q4 **Algebraic method**: There are various ways of setting out. This is a compact one:

$T = 2\pi\sqrt{\dfrac{l}{g}}$, so $g_{moon} = g_{Earth} \times \left(\dfrac{T_{Earth}}{T_{moon}}\right)^2 = 9.81 \times \left(\dfrac{100}{240}\right)^2 = 1.70 \text{ m s}^{-2}$ (3 sf)

Alternative method: Calculate length of pendulum on Earth ⟶ 0.248 m. Then use this to calculate g on Moon from the period ⟶ 1.70 m s^{-2}.

Q5 (a) Time taken $= \frac{1}{2} \times$ period $= 0.5 \times 2\pi\sqrt{\dfrac{m}{k}} = 0.5 \times 2\pi\sqrt{\dfrac{0.200 \text{ kg}}{40 \text{ N m}^{-1}}} = 0.22 \text{ s}$ (2 sf)

(b) It is true that there is a larger resultant upward force at the maximum extension but there is a larger distance to travel. The period of a body undergoing SHM is independent of the amplitude, e.g. for a mass on a spring it depends only on the mass and the stiffness: $T = 2\pi\sqrt{m/k}$, and the time involved is half the period, so Fergus's conclusion is invalid.

Q6 The spring constant $k = \dfrac{mg}{l}$, where m is the mass of object, so $\dfrac{m}{k} = \dfrac{l}{g}$.

The period, T, of oscillation of the object on the spring is given by $T = 2\pi\sqrt{\dfrac{m}{k}}$.

$\therefore$ Substituting for $\dfrac{m}{k}$ ⟶ $T = 2\pi\sqrt{\dfrac{l}{g}}$. But this is just the same as the period of a simple pendulum of length l, so it is not a fluke and Davinder is (on this occasion) incorrect.

Q7 (a) If a pendulum of length l is displaced by angle θ, the distance d of the bob below the pivot is given by:

$d = l\cos\theta$.

So the height raised, $h = l\,(1 - \cos\theta)$

$\therefore$ Gain in GPE $= mg\,l\,(1 - \cos\theta)$

$= 0.100 \times 9.81 \times 1.00\,(1 - \cos 11.5°)$

$= 0.0197 \text{ J} \sim 0.02 \text{ J}$

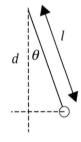

(b) (i)

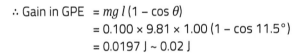

(ii)

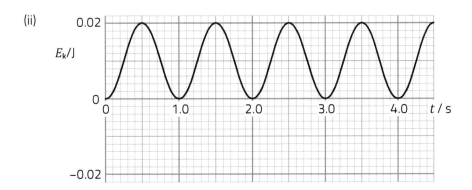

Q8 (a) Air resistance will act on the mass and disc. This is a resistive force and always opposes the motion. Energy will be lost to random kinetic energy of air molecules, i.e. internal energy of the air will increase slightly.

(b) From 0.15 s to 1.95 s the amplitude halves (from 0.2 m s^{-1} to 0.1 m s^{-1}), i.e. a half-life of 1.80 s. Also halves (0.1 to 0.05) from 1.95 s to 3.75 s. Again a half-life of 1.80 s.
Final check: another half-life is 0.75 s to 2.55 s (0.16 to 0.08). Another half-life of 1.80 s.
Constant half-life means an exponential decay and Sophie is therefore correct.

(c) Velocity halves so the KE will reduce to one quarter, since KE = $\frac{1}{2}$ $mv2$
Hence 75% of the KE has been lost or dissipated.

Q9 (a) Critical damping is when the resistive forces are just large enough to stop oscillations occurring.

(b) Car suspension. If oscillations took place, the tyres could leave the road, resulting in less friction and increased braking distances. This would increase the number of accidents occurring (this can have an enormous effect on a bumpy road).

Q10 (a) Forced oscillations is when a periodic driving force acts on an oscillatory system. The system then oscillates at the frequency of the driving force.

(b) (i) It is the periodic (or sinusoidal) force applied to it as a 'push and pull' by the vibrating pin via the weak spring.

(ii) From the graph, the resonant frequency is 0.8 Hz and this must be the natural frequency of the pendulum. Even though there is a fair amount of damping, the frequency will only decrease a very tiny percentage away from the natural frequency.

Period, $T = \dfrac{1}{0.80\ \text{Hz}}$ = 1.25 s.

Rearranging $T = 2\pi\sqrt{\dfrac{l}{g}} \longrightarrow l = \left(\dfrac{T}{2\pi}\right)^{2} g = \left(\dfrac{1.25}{2\pi}\right)^{2} \times 9.81 = 0.39$ m (2 sf)

Section 3.3: Kinetic theory

Q1 Gases consist of a large number of molecules in random motion in (otherwise) empty space.
The volume of the molecules themselves is a negligible fraction of the volume of the gas.
Collisions between molecules are perfectly elastic and take a negligible time.
The molecules exert negligible forces on one another except during collision.

Q2 The mole is the amount of a substance with as many particles as there are in exactly 12 g of carbon-12.
The Avogadro constant is the number of particles per mole [$\sim 6.02 \times 10^{23}$ mol^{-1}]

[Note that these are the definitions from the Terms and Definitions booklet. They are out of date. Since 2019 the mole has been defined as exactly $6.022\ 140\ 76 \times 10^{23}$ particles and hence the Avogadro constant, N_A, is $6.022\ 140\ 76 \times 10^{23}$ mol^{-1}. Either pair of definitions will be credited in WJEC examinations.]

Q3 The molecules of a gas, in their random rapid motion, collide with the walls of the container. In such a collision, a molecule suffers a change in momentum inwards at right angles (on the average) to the wall. There are a large number of such collisions per second on any area of wall, so the wall exerts a force on the molecules (equal to the change in momentum per second – Newton's second law). By Newton's third law, the molecules exert an equal and opposite force on the wall. The pressure is the magnitude of this force divided by the area of the wall.

If the temperature is increased, the molecular speeds increase, so there are more collisions per second with the wall and each of them results in a larger change of momentum. Hence, the pressure exerted by the gas molecules increases with temperature.

Q4 The ideal gas equation of state is $pV = nRT$, where n is the amount of the gas, ie. the number of moles. The equation can also be expressed as $pV = NkT$, where N is the number of molecules of the gas.

$N = nN_A$, so the second equation can be written $pV = nN_A kT$.

Comparing this with the first equation, $nN_A k = nR$. Hence $k = \dfrac{R}{N_A}$.

Q5 For a single molecule, KE $= \frac{1}{2} mc^2$, so the mean molecular kinetic energy $= \frac{1}{2} m\overline{c^2}$ and the total kinetic energy in a gas, $U = \frac{1}{2} Nm\overline{c^2}$, where N is the number of molecules.

Equating the right-hand sides of the given equations: $\frac{1}{3} Nm\overline{c^2} = nRT$

But $\frac{1}{3} Nm\overline{c^2} = \frac{2}{3} U$, $\therefore \frac{2}{3} U = nRT$, and hence $U = \frac{3}{2} nRT$

[Note: this expression is only valid for the translational kinetic energy of molecules – it doesn't include any contribution from rotational kinetic energy which is significant in gases with polyatomic molecules.]

Q6 $pV = \frac{1}{3} Nm\overline{c^2}$

so $V = \dfrac{Nm\overline{c^2}}{3p} = \dfrac{\text{mass of gas} \times \overline{c^2}}{3p}$

$= \dfrac{3.0 \,\text{mol} \times 0.028 \,\text{kg mol}^{-1} \times (550 \,\text{m s}^{-1})^2}{3 \times 140 \times 10^3 \,\text{Pa}}$

$= 0.061 \,\text{m}^3$

Q7 (a) $pV = nRT$

so $n = \dfrac{pV}{RT} = \dfrac{102 \times 10^3 \,\text{Pa} \times 0.89 \,\text{m}^3}{8.31 \,\text{J mol}^{-1} \,\text{K}^{-1} \times 298 \,\text{K}} = 36.7 \,\text{mol}$, that is 37 mol (to 2 sf)

(b) $\frac{1}{2} m\overline{c^2} = \frac{3}{2} kT$

so $c_{\text{rms}} = \sqrt{\overline{c^2}} = \sqrt{\dfrac{3kT}{m}} = \sqrt{\dfrac{3 \times 1.38 \times 10^{-23} \,\text{J K}^{-1} \times 298 \,\text{K}}{6.64 \times 10^{-27} \,\text{kg}}} = 1\,360 \,\text{m s}^{-1}$

(c) $pV = nRT$

so $V = \dfrac{nRT}{p} = \dfrac{36.7 \,\text{mol} \times 8.31 \,\text{J mol}^{-1} \,\text{K}^{-1} \times 232 \,\text{K}}{23 \times 10^3 \,\text{Pa}} = 3.1 \,\text{m}^3$

We are assuming that (i) the inward pressure exerted by the stretched balloon skin itself is negligible compared with the 23 kPa atmospheric pressure, (ii) all the helium has reached 232 K, (iii) the helium is behaving as an ideal gas. [Note: the examiner would expect one assumption and anticipate the first of these.]

Q8 (a) The total number of moles stays constant. Calculating it from the initial data

$n = \left(\dfrac{pV}{RT}\right)_{\text{left}} + \left(\dfrac{pV}{RT}\right)_{\text{right}}$

$= \dfrac{1.02 \times 10^5 \,\text{Pa} \times 37.0 \times 10^{-3} \,\text{m}^3}{8.31 \,\text{J mol}^{-1} \times 293 \,\text{K}} + \dfrac{6.50 \times 10^5 \,\text{Pa} \times 22.5 \times 10^{-3} \,\text{m}^3}{8.31 \,\text{J mol}^{-1} \times 293 \,\text{K}}$

$= 1.55 \,\text{mol} + 6.00 \,\text{mol} = 7.55 \,\text{mol}$

After thermal equilibrium, this number of moles occupies $59.5 \times 10^{-3} \,\text{m}^3$ at 293 K,

$$\text{so } p = \frac{nRT}{V} = \frac{7.55 \text{ mol} \times 8.31 \text{ J mol}^{-1} \text{ K}^{-1} \times 293 \text{ K}}{59.5 \times 10^{-3} \text{ m}^3}$$

$$= 3.1 \times 10^5 \text{ Pa}$$

(b) [Assuming opening of tap is equivalent to release of light gas-tight piston separating the two containers and that the process is quick enough for heat flow to be negligible.] High pressure gas on the right will do work on the gas on the left by compressing it, lose internal energy and cool; low pressure gas on the left will have work done on it and become warmer. So Tudor is correct.

Q9 (a) $p = \frac{1}{3}\rho\overline{c^2}$

$$\text{so } c_{rms} = \sqrt{\overline{c^2}} = \sqrt{\frac{3p}{\rho}} = \sqrt{\frac{3 \times 112 \times 10^3 \text{ Pa}}{1.35 \text{ kg m}^{-3}}} = 499 \text{ m s}^{-1}$$

(b) Mass of gas = density × volume = ρV

$$\text{But } n = \frac{pV}{RT}$$

$$\text{So mass per mole} = \frac{\rho V}{n} = \frac{\rho RT}{p} = \frac{1.35 \text{ kg m}^{-3} \times 8.31 \text{ J mol}^{-1} \text{ K}^{-1} \times 293 \text{ K}}{112 \times 10^3 \text{ Pa}}$$

$$= 0.0293 \text{ kg mol}^{-1}$$

(c) (i) $n = \frac{pV}{RT} = \frac{935 \times 10^3 \text{ Pa} \times 1.5 \times 10^{-3} \text{ m}^3}{8.31 \text{ J mol}^{-1} \text{ K}^{-1} \times 320 \text{ K}} = 0.527 \text{ mol}$

So mass of air in bottle $= 0.527 \times 10^{-3} \text{ mol} \times 0.0293 \text{ kg mol}^{-1}$

$= 0.015 \text{ kg}$

Assumption: the air is an ideal gas.

(ii) Work is being done on the gas as it is pumped in. Not much heat escapes, so the internal energy of the gas rises, hence its temperature increases.

Section 3.4: Thermal physics

Q1 The internal energy of a system is the sum of the kinetic and potential energies of the particles in the system.

Q2 At absolute zero, the internal energy of a system is the minimum possible. It is impossible to extract energy from the sy stem.

Q3 (a) Unlike in other systems, the molecules of an ideal gas do not exert forces on each other, so there is no potential energy component to the internal energy – it is entirely kinetic energy.

(b) $U = \frac{3}{2}NkT$. $N = N_A \times \frac{30 \text{ g}}{20 \text{ g}} = 1.5N_A$, $T = (273.15 + 26.85) \text{ K} = 300 \text{ K}$

$\therefore U = \frac{3}{2} \times 1.5 \times 6.022 \times 10^{23} \times 1.38 \times 10^{-23} \text{ J K}^{-1} \times 300 \text{ K} = 5600 \text{ J (2 sf)}$

Q4 Heat is the flow of energy from one system to another due to a temperature difference. The flow is into the system with a lower temperature.

Q5 Thermal equilibrium means that there is no heat flow between the systems – their temperatures are the same.

Q6 (a) ΔU = the increase in internal energy of the system
Q = the heat flow into the system
W = the work done by the system
Q leads to a gain in internal energy of the system and W a loss. The net gain in internal energy (ΔU) is equal to the energy input due to heat minus the energy loss due to work.

(b) $W = p\Delta V$ where p is the pressure and ΔV the change in volume of the system. The volume of a solid or a liquid is almost constant, so $\Delta V \sim 0$. Hence $W = 0$.

Q7 The specific heat capacity of a substance is the heat required to raise the temperature of the substance, per unit mass per unit temperature rise.

Q8 The apparatus is set up without the Bunsen burner and with cold water and ice (at $0\,°C$) in the beaker. The temperature is measured using the thermometer and the pressure of the air in the flask measured using the pressure gauge. The temperature of the water is raised using the Bunsen in a series of steps of approximately $10\,°C$ up to $100\,°C$. At each step the Bunsen is removed, the water stirred, and time allowed for the system to equilibrate (stirring will help); the pressure of the air column and the temperature are measured.

The pressure of the air is plotted against the temperature (in $°C$) and a best fit straight line drawn. The intercept of this line on the temperature axis is the estimate of absolute zero.

Q9 First law of thermodynamics: $\Delta U = Q - W$.
$Q = 0$ so $\Delta U = -W$. The gas expands, so $W > 0$. Hence $\Delta U < 0$.
$U = \frac{3}{2}NkT$. N and k are constant so, if $\Delta U < 0$, $\Delta T < 0$.

Q10 (a) $W = p\Delta V = 1.42 \times 10^5\,Pa \times 2.7 \times 10^{-3}\,m^3$
$= 380\,J$ (2 sf)

(b) $W = p\Delta V = -380\,J + 1.42 \times 10^5\,Pa \times (-1.5 \times 10^{-3}\,m^3)$
$= -590\,J$ (2 sf)

Q11 Assumptions: Heat exchange with the environment is negligible; the heat capacity of the container is negligible. In this case:

Loss of internal energy of water = Gain of internal energy of carrots

Let θ = equilibrium temperature:

$\therefore 1.2\,kg \times 4210\,J\,kg^{-1}°C^{-1} \times (100\,°C - \theta) = 0.700\,kg \times 1880\,J\,kg^{-1}°C^{-1} \times (\theta - 20\,°C)$
[Omitting the units for clarity]

$\therefore 505\,200 - 5052\theta = 1316\theta - 26\,320$

$\therefore 6368\theta = 531\,520$

$\therefore \theta = \dfrac{531\,520}{6368}\,°C = 83\,°C$ (2 sf)

Q12 (a) (i) 0 [because the volume is constant]

(ii) $W = p\Delta V = 0.85 \times 10^5\,Pa \times (7.1 - 16.7) \times 10^{-3}\,m$
$= -816\,J$

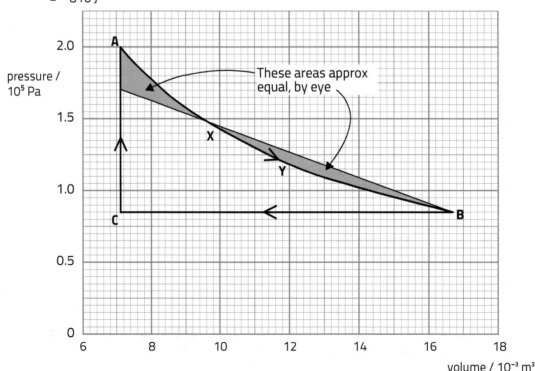

(iii) W = Area under graph A⟶C = Area under estimated straight line (see diagram)

= area of a trapezium

= $\frac{1}{2} \times (1.70 + 0.85) \times 10^5 \times (16.7 - 7.1) \times 10^{-3}$

= 1224 J

(b) If T is constant, pV is a constant:

[Omitting the factor $10^5 \times 10^{-3} = 100$ as this is only a comparison]

Checking values: $(pV)_A = 2.0 \times 7.1 = 14.2$; $(pV)_B = 0.85 \times 16.7 = 14.2$

$(pV)_X = 1.42 \times 10 = 14.2$; $(pV)_Y = 1.18 \times 12 = 14.2$

Hence the values of pV are the same at these four places suggesting that the temperature is constant along AB.

(c)

	AB	BC	CA	ABCA
ΔU / J	0	−1225	1225	0
Q / J	1200	−2045	1225	380
W / J	(a)(iii) 1200	(a)(ii) −820	(a)(i) 0	380

Some calculations: Using $U = \frac{3}{2} pV$, $U_A = U_B = \frac{3}{2} \times 14.2 \times 10^2$ J = 2130 J;

$U_C = \frac{3}{2} \times 0.85 \times 10^5$ Pa $\times 7.1 \times 10^{-3}$ m^3 = 905 J

∴BC: $\Delta U = 905 - 2130 = -1225$ J, ∴ $Q = \Delta U + W = -1225 - 820 = -2045$ J

and CA: $(\Delta U)_{CA} = -(\Delta U)_{BC} = 1225$ J, ∴ $Q = \Delta U + W = 1225 + 0 = 1225$ J

Q13 Calculations: Heat supplied each minute = 12.00 V $\times$ 4.20 A $\times$ 60 s = 3024 J

∴ Temperature rise per minute = $\dfrac{Q}{mc} = \dfrac{3024\ \text{J}}{1.00\ \text{kg} \times 900\ \text{J kg}^{-1}\,{}^\circ\text{C}^{-1}} = 3.36\ ^\circ\text{C}$

Suppose heat only starts reaching the thermometer by conduction after about 1 minute [Note: you wouldn't be penalised assuming the graph started rising straight away]

⟶ Potential temperature rise in 20 min = $3.36 \times 19 = 63.8\ ^\circ\text{C}$ ⟶ top temp = 84 °C

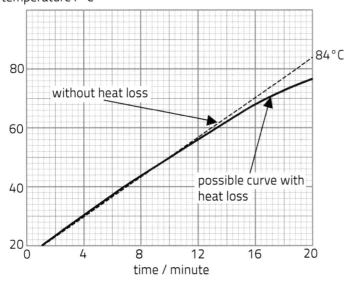

temperature /°C

without heat loss

possible curve with heat loss

84 °C

time / minute

Q14 (a) $[p\,\Delta V] = [p][\Delta V] = (\text{N m}^{-2})\ \text{m}^3 = \text{N m} = \text{J} = [W]$ QED

(b) (i) **Method 1**: (calculating n)

Initially, $n = \dfrac{pV}{RT} = \dfrac{100 \times 10^3\ \text{Pa} \times 110 \times 10^{-6}\ \text{m}^3}{8.31\ \text{J mol}^{-1}\,\text{K}^{-1} \times 293\,\text{K}} = 4.52$ mmol (to 3 sf)

Finally, $n = \dfrac{pV}{RT} = \dfrac{100 \times 10^3\ \text{Pa} \times 140 \times 10^{-6}\ \text{m}^3}{8.31\ \text{J mol}^{-1}\,\text{K}^{-1} \times 373\,\text{K}} = 4.52$ mmol (to 3 sf)

The final amount of gas is the same as the initial to 3 sf, so a negligible amount has escaped.

Method 2: (not calculating n)

If the amount of gas and the pressure are constant, $V \propto T$:

$$\left(\frac{V}{T}\right)_{\text{initial}} = \frac{110 \times 10^{-6}\, \text{m}^3}{293\, \text{K}} = 3.75 \times 10^{-7}\, \text{m}^3\, \text{K}^{-1} \text{ (to 3 sf)}$$

$$\left(\frac{V}{T}\right)_{\text{final}} = \frac{140 \times 10^{-6}\, \text{m}^3}{373\, \text{K}} = 3.75 \times 10^{-7}\, \text{m}^3\, \text{K}^{-1} \text{ (to 3 sf)}$$

These are the same to 3 sf, so the amount escaping is negligible.

(ii) The work done by the gas, $W = p\Delta V = 100 \times 10^3\, \text{Pa} \times (140 - 110) \times 10^{-6}\, \text{m}^3 = 3.0\, \text{J}$

$\Delta U = \frac{3}{2} \times nR\Delta T = \frac{3}{2} \times 4.52 \times 10^{-3}\, \text{mol} \times 8.31\, \text{J mol}^{-1}\, \text{K}^{-1} \times (100 - 20)\, \text{K}$

$= 4.50\, \text{J}$

[Or: $\Delta U = \frac{3}{2} \times nR\Delta T = \Delta U = \frac{3}{2} \times p\Delta V$ (because p is constant) $= \frac{3}{2} W = 4.5\, \text{J}$]

From the first law of thermodynamics: $Q = \Delta U + W = 3.0\, \text{J} + 4.5\, \text{J} = 7.5\, \text{J}$

(iii) From the above numbers:

Heat needed to raise the temperature / mol and / degree $= \dfrac{7.5\, \text{J}}{4.52 \times 10^{-3}\, \text{mol} \times 80\, \text{K}}$

$= 21\, \text{J mol}^{-1}\, \text{K}^{-1}$

This is in agreement with Lucia's statement. However, if the gas were kept at a fixed volume, no work would be done by the gas, so the heat needed to raise the internal energy by the same amount would be less (because the internal energy of an ideal gas depends only on its temperature). So the heat needed isn't always 21 J.

Section 3.5: Nuclear decay

Q1 The β particles could be <u>positrons</u> (β⁺ radiation). The particles are of <u>high energy</u> and originate in the <u>unstable nuclei</u> of certain atoms.

Q2 (a) The activity of a radioactive nuclide is the number of disintegrations per unit time.
Unit: becquerel (Bq) which is equivalent to s^{-1}.

(b) 1 year $= 60 \times 60 \times 24 \times 365\, \text{s} = 3.15 \times 10^7\, \text{s}$

If there are N atom, no. of decays in 1 year $= 9.11 \times 10^{-13} \times 3.15 \times 10^7\, N$

$= 2.87 \times 10^{-5}\, N$

∴ Probability of a particular atom decaying $= 2.87 \times 10^{-5} = \dfrac{1}{34\,800}$

∴ The true probability is 1 in 34 700 which is indeed less than 1 in 30 000.

Q3 (a) All the readings are the same to within the expected variability of the random emissions. Any α radiation from the sample would not penetrate 10 cm of air, so the experiment gives no information about α. The lack of a reduction with the thin aluminium suggests that the sample is a γ emitter.

(b) A background reading should be taken to establish the readings from the source alone. The readings should be repeated at 2 cm distance to allow any α emissions to be detected.

Q4 (a) $^{235}_{92}\text{U} \rightarrow\ ^{231}_{90}\text{Th} +\ ^{4}_{2}\text{He}$ [or $^{4}_{2}\alpha$]

(b) (i) The mass number (nucleon number), A, must be $235 - 4n$ where n is an integer because the $^{4}_{2}\text{He}$ has $A = 4$. $207 = 235 - 4 \times 7$, so $^{207}_{82}\text{Pb}$ is the end of the decay series.

(ii) $206 = 238 - 4 \times 8$, so there are 8 alpha decays. Without β⁻ this would produce a nuclide with $Z = 92 - 2 \times 8 = 76$. So 6 β⁻¹ decays are needed to give $Z = 82$.

$$^{238}_{92}\text{U} \rightarrow\ ^{206}_{82}\text{Th} + 8\,^{4}_{2}\text{He} + 6\,^{0}_{-1}\text{e}$$

(iii) There is no whole number n for which $233 - 4n$ is equal to 206, 207 or 208, so any isotope of lead produced, e.g. 205, would be unstable and cannot be the end product.

Q5

Method 1: equal times → equal ratios

(a) (i) After 1 year $A \rightarrow \dfrac{9.76}{11.50} A = 0.849A$.

∴ In another year, the count rate will be $0.849 \times 9.76 = 8.28$

(ii) Count rate $= (0.849)^{10} \times 9.76 = 1.90$ count per second.

(b) In n years, count rate $= 9.76(0.849)^n$.

If $9.76(0.849)^n = 0.42$, taking logs: $\ln 9.76 + n \ln 0.849 = \ln 0.42$

∴ $n = \dfrac{\ln 0.42 - \ln 9.76}{\ln 0.849} = 19.2$ years (3 sf)

Method 2: calculating the decay constant

(a) (i) After 1.00 year, $9.76 = 11.50e^{-1.00\lambda}$, so $\ln 9.76 = \ln 11.50 - \lambda \rightarrow \lambda = 0.164$ year^{-1}

∴ In another year, the $C = 9.76e^{-0.164 \times 1.00} = 8.28$

(ii) $C = 9.76e^{-0.164 \times 10.00} = 1.89$ count per second.

(b) In n years $C = 9.76e^{-0.164n}$.

If $9.76e^{-0.164n} = 0.42$, taking logs: $\ln 9.76 - 0.164n = \ln 0.42$

∴ $n = \dfrac{\ln 9.76 - \ln 0.42}{0.164} = 19.2$ years (3 sf)

Method 3: Using half-lives $C = C_0\, 2^{-n}$ [left for you]

Q6 (a) $^{1}_{0}n + {}^{14}_{7}N \rightarrow {}^{14}_{6}C + {}^{1}_{1}H$

(b) $^{14}_{6}C \rightarrow {}^{14}_{7}N + {}^{0}_{-1}e + {}^{0}_{0}\overline{\nu}_e$

(c) (i) $\lambda = \dfrac{\ln 2}{5730 \text{ year}} = 1.210 \times 10^{-4}$ year^{-1} $[= 3.83 \times 10^{-12}$ s$^{-1}]$

(ii) Minimum age: $e^{-\lambda t} = \dfrac{0.853}{1.250} = 0.6824$, ∴ $-1.210 \times 10^{-4} t = \ln 0.6824$

∴ $t_{min} = 3158$ year.

Maximum age: $e^{-\lambda t} = \dfrac{0.849}{1.250} = 0.6792$, ∴ $-1.210 \times 10^{-4} t = \ln 0.6792$

∴ $t_{max} = 3197$ year. ∴ Age $= 3180 \pm 20$ year

(iii) **Qualitative answer:** The fraction of ^{14}C in modern objects is less that it would be without the addition of the fossil fuels. Objects which are old also have a lower fraction of ^{14}C, which is the same effect, so Sioned is correct.

Quantitative answer: If apparent age of modern object $= t$, $e^{-\lambda t} = 0.97$

∴ $t = \dfrac{\ln 0.97}{-1.21 \times 10^{-4}} = 250$ years. So modern objects appear 250 years old and Sioned is correct.

Q7 (a) With 8 faces, probability of remaining after 1 throw $= \dfrac{7}{8} = 0.875$.

∴ No. remaining $= 800 \times 0.875$ ∴ After n throws $800 \times (0.875)^n$

(b) For half to remain, $(0.875)^n = 0.5$,

∴ $n = \dfrac{\ln 0.5}{\ln 0.875} = 5.19$ throws. The points agree with the theoretical curve and half-life.

(c) Theoretical no. remaining = $800 \times (0.75)^n \rightarrow$ In 2-throw steps: 450, 253, 142, 80, 45, 25

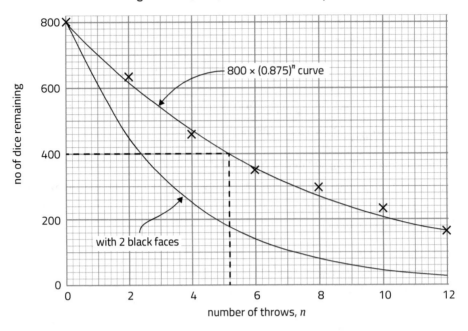

Q8 (a) Range of 1.0 MeV β particles = 0.41 cm

∴ Range in glass = $\dfrac{1.0 \times 10^3 \text{ kg m}^{-3}}{2.5 \times 10^3 \text{ kg m}^{-3}} \times 4.1$ cm = 1.6(4) cm

(b) The range of 2.0 MeV β particles is 0.97 cm, showing that they lose the first MeV in 0.56 cm (i.e. 0.97 cm – 0.41 cm) and the last MeV in 0.41 cm, hence Dylan is correct.

(c) (i) $^6_3\text{Li} + ^1_0\text{n} \rightarrow ^3_1\text{H} + ^4_2\text{He}$

(ii) $^3_1\text{H} \rightarrow ^3_2\text{He} + ^0_{-1}\text{e} + ^0_0\overline{v}_e$

(iii) The range of 0.1 MeV β particles is about 0.01 cm in water and therefore about 0.04 mm in glass. Hence, the β particles cannot penetrate the walls of the tubes.

Q9 (a)

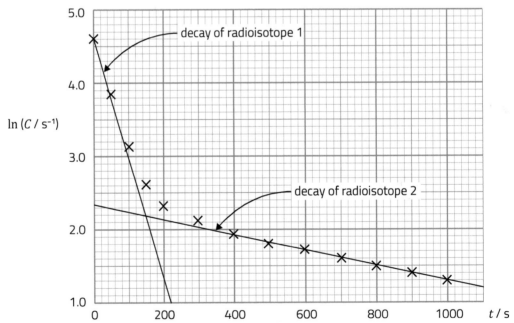

The ln C vs. t graph consists of two straight line portions of different negative gradients. [Note: The curved part represents the period in which both isotopes contribute similarly to the counts.]

(b) (i) Decay constant = – gradient = $-\dfrac{1.19 - 2.33}{1100 \text{ s}}$

∴ $\lambda = 1.04 \times 10^{-3}$ s^{-1}

(ii) For isotope 2, at $t = 0$, $\ln(C/\text{s}^{-1}) = 2.33$, where C is the count rate

$\therefore C/\text{s}^{-1} = e^{2.33} = 10.3$

i.e. $C = 10.3 \text{ s}^{-1}$

(c) Initial total count rate $= e^{4.6} = 99.5 \text{ s}^{-1}$

$\therefore$ Count rate from isotope 1 $= 99.5 - 10.3 = 89.2 \text{ s}^{-1}$

Q10 (a) $^{238}_{92}\text{U} \rightarrow {}^{234}_{90}\text{Th} + {}^{4}_{2}\text{He}$

$^{234}_{90}\text{Th} \rightarrow {}^{234}_{91}\text{Pa} + {}^{0}_{-1}\text{e} + {}^{0}_{0}\overline{\nu}_e$

(b) After n half-lives $C = C_0 \times 2^{-n}$, $\therefore n = \dfrac{\ln(C_0/C)}{\ln 2}$

The maximum decay is $492 \rightarrow 77$ and the minimum decay is $448 \rightarrow 95$ counts

In 3 minutes: $n_{max} = \dfrac{\ln(492/77)}{\ln 2} = 2.68$,

$\therefore t_{1/2 \text{ min}} = \dfrac{3.0 \text{ minutes}}{2.68} = 1.12 \text{ minutes}$

and $n_{min} = \dfrac{\ln(448/95)}{\ln 2} = 2.23$,

$\therefore t_{1/2 \text{ max}} = \dfrac{3.0 \text{ minutes}}{2.23} = 1.34 \text{ minutes}$

The figure of 1.17 minutes is within the experimental range, so the student's results are consistent with it.

(c) Repeat the experiment several times and measure the 10 second count rates C, more frequently, e.g. every 20 s. Add the results for each time together and plot a graph of $\ln C$ against t. The straight-line graph has a gradient of $-\lambda$ from which the half-life can be calculated using $t_{1/2} = (\ln 2)/\lambda$.

Q11 (a) A charged particle, q, moving with a velocity, v, across a magnetic field, B, experiences a force, F, at right angles to B and v in the direction given by Fleming's Left Hand Motor rule. If q is negative, the direction of F is opposite. In this case, the charges are accelerated downwards in the field showing them to be negative (β^-).

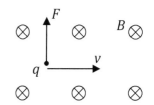

(b) Photons are uncharged and so will be undeflected, i.e. they pass through in a straight line. α particles are positively charged, so experience an upward force; however, they have a much greater mass than β particles so the deflection is very small and would not be observed with this simple arrangement.

Section 3.6: Nuclear energy

Q1 (a) The 'energy to hold the particles together' would have mass, given by $E = mc^2$. So the total mass would be the mass of the protons + the mass of the neutrons + the mass of this energy.

(b) Binding energy is the energy required to separate the component nucleons.

Q2 13.6 eV $= 2.18 \times 10^{-18}$ J.

This energy has mass $= \dfrac{2.18 \times 10^{-18} \text{ J}}{(3.00 \times 10^8 \text{ m s}^{-1})^2} = 2.42 \times 10^{-35} \text{ kg} = 1.46 \times 10^{-8} \text{ u}$.

So the mass of the atom is 0.000 000 015 u less than the sum of the masses of the proton and the electron. Hence they are different to this number of significant figures.

Q3 (a) Mass deficit $= 2 \times (1.007\,276 + 1.008\,665 + 0.000\,549) - 4.002\,604$ u

$= 0.030\,376$ u

$\therefore$ Binding energy $= 0.030\,376 \times 931 \text{ MeV} = 28.3 \text{ MeV}$

(b) Binding energy per nucleon $= \dfrac{28.3}{4} = 7.07 \text{ MeV nuc}^{-1}$

Q4 (a) Power $= 4\pi (1.50 \times 10^{11} \text{ m})^2 \times 1370 \text{ W m}^{-2}$

$= 3.87 \times 10^{26} \text{ W}$

(b) Rate of mass loss $= \dfrac{3.87 \times 10^{26} \text{ W}}{(3.00 \times 10^8 \text{ m s}^{-1})^2} = 4.30 \times 10^9 \text{ kg s}^{-1} = 4.3$ million tonnes per second

Q5 (a) Decay constant of $^{235}_{92}U = \dfrac{\ln 2}{7.1 \times 10^8} = 9.76 \times 10^{-10}$ year^{-1} = 3.09×10^{-17} s^{-1}.

No. of atoms of $^{235}_{92}U = \dfrac{1}{0.235} \times 6.02 \times 10^{23} = 2.56 \times 10^{24}$

∴ Activity = $N\lambda = 7.92 \times 10^7$ Bq

(b) (i) Energy is released by each decay. Nearly all the alpha particles are absorbed in the lump. This increases the vibrational energy of the $^{235}_{92}U$ atoms, increasing the temperature.

(ii) Mass loss per decay = 235.043 930 – (231.036 304 + 4.002 604) u

$= 0.005\ 022$ u

∴ Energy per decay = 4.68 MeV = 7.48×10^{-13} J

∴ Total power = 7.48×10^{-13} J $\times 7.92 \times 10^7$ Bq

$= 5.9 \times 10^{-5}$ W

This is undetectably small, so Michael is correct.

Q6 (a) $^6_3\text{Li} + ^1_0\text{n} \rightarrow ^3_1\text{H} + ^4_2\text{He}$

(b) $^3_1\text{H} \rightarrow ^3_2\text{He} + ^0_{-1}\text{e} + ^0_0\overline{\nu}_e$

(c) The reaction is $^3_1\text{H} + ^2_1\text{H} \rightarrow ^4_2\text{He} + ^1_0\text{n}$

Binding energy of ^4_2He = 4 × 7.1 MeV = 28.4 MeV

Binding energy of $^3_1\text{H} + ^2_1\text{H}$ = 3 × 2.8 + 2 × 1.1 = 10.6 MeV

∴ Energy released per reaction = 28.4 MeV - 10.6 MeV = 17.8 MeV

Appropriate mixture = 600 g ^3_1H + 400 g ^2_1H, i.e. 200 moles of reactants

∴ Energy release = $6.02 \times 10^{23} \times 200 \times 17.8$ MeV $\times 1.60 \times 10^{-13}$ J MeV^{-1}

$= 3.4 \times 10^{14}$ J

Q7 (a) Mass loss = 4 × 1.007 825 – 4.002 604 – 2 × 0.000549 = 0.027 598 u

∴ Energy release = 0.027 598 u × 931 MeV u^{-1}

$= 25.694$ MeV = 25.7 MeV (3 sf)

(b) Mass loss = 3 × 4.002 604 u – 12 u (exactly) = 0.007 812 u

∴ Energy release = 7.27 MeV

(c) (i) Luminosity = power emitted $\propto r^2 T^4$.

∴ $L = 10^2 \times 0.9^4 = 65.6 \times$ current value.

(ii) Energy release per triple α reaction = $\dfrac{7.27}{25.7} = 0.283$ of that in the hydrogen fusion.

Three hydrogen fusions are needed for each triple alpha so only one-third of the reactions can take place.

∴ Energy released in triple alpha period = $\dfrac{7.27}{3 \times 25.7} = 0.094 \times$ H fusion period.

∴ Lifetime $\sim \dfrac{0.094}{65.6} \times 9 \times 10^9$ years = 13 million years.

Q8 $^{56}_{26}\text{Fe} + 179^1_0\text{n} \longrightarrow ^{235}_{92}\text{U} + 66^0_{-1}\text{e} + 66^0_0\overline{\nu}_e$

179 neutrons are needed to balance the mass numbers: 56 + 179 = 235

66 electrons are needed to balance the atomic numbers: 26 = 92 + (–66)

Lepton number is conserved, so for each electron there is one anti-neutrino.

Q9 (a) The mass of two ^4_2He atoms = 8.005 208 u. This is less than the mass of a ^8_4Be atom. All the nuclear conservation laws are conserved so it does not involve the weak interaction.

(b) (i) Momentum is conserved so if the momenta of the two daughter ^4_2He nuclei must add to zero, i.e. they must be equal and opposite.

(ii) Mass loss = 8.005 305 – 8.005 208 u = 0.000 097 u

∴ KE release = 0.000 097 × 931 = 0.090 307 MeV = 1.44×10^{-14} J

∴ Energy of each nucleus = 0.72×10^{-14} J

∴ $7.2 \times 10^{-15} = \dfrac{1}{2} \times 4.0 \times 1.66 \times 10^{-27} v^2$

∴ $v = 1.5 \times 10^6$ m s^{-1}

Practice questions: Unit 4: Fields and Options

Section 4.1: Capacitance

Q1 $Q = CV = 22$ mF $\times$ 12 V $= 260$ mC (2 sf)
So the charges on the plates are $+260$ mC and -260 mC.

Q2 (a) $C = \dfrac{\varepsilon_0 A}{d}$, so $d = \dfrac{\varepsilon_0 A}{C} = \dfrac{8.85 \times 10^{-12}\ \text{F m}^{-1} \times (0.10\ \text{m})^2}{500 \times 10^{-12}\ \text{F}} = 1.8 \times 10^{-4}$ m (0.18 mm)

(b) Putting a polymer between the plates of a capacitor increases its capacitance. The capacitance is inversely proportional to the plate separation. Hence, to achieve the same capacitance, the plate separation needs to be bigger.

Q3 (a) $C = \dfrac{\varepsilon_0 A}{d} = \dfrac{8.85 \times 10^{-12}\ \text{F m}^{-1} \times 64 \times 10^{-4}\ \text{m}^2}{0.40 \times 10^{-3}\ \text{m}} = 1.42 \times 10^{-10}$ F

$Q = CV = 1.42 \times 10^{-10}$ F $\times$ 30 V $= 4.2 \times 10^{-9}$ C (2 sf)

(b) $U = \frac{1}{2}CV^2 = 0.5 \times 1.42 \times 10^{-10}$ F $\times (30\ \text{V})^2 = 6.4 \times 10^{-8}$ J

(c) $E = \dfrac{V}{d} = \dfrac{30\ V}{0.40 \times 10^{-3}\ \text{m}} = 75\,000$ V m^{-1}

Q4 (a) $U = \frac{1}{2}\dfrac{Q^2}{C}$. As Q is constant $U \propto C^{-1}$.

But $C = \dfrac{\varepsilon_0 A}{d}$, so $C \propto d^{-1}$ and so $U \propto d$. Hence the energy is doubled.

(b) The opposite charges attract, so work has to be done to separate them. The energy has come from the agency which separated the plates.

Q5 $U = \frac{1}{2}CV^2 = \frac{1}{2}\dfrac{\varepsilon_0 A}{d}V^2$

But $E = \dfrac{V}{d}$, so $V = Ed$ and substituting for V gives:

$U = \frac{1}{2}\dfrac{\varepsilon_0 A}{d}(Ed)^2$ which simplifies to $U = \frac{1}{2}\varepsilon_0 E^2 \times (Ad) = \frac{1}{2}\varepsilon_0 E^2 \times$ volume

Q6 Total capacitance, $C = \dfrac{C_1 C_2}{C_1 + C_2} = \dfrac{7.0 \times 3.0}{7.0 + 3.0}$ μF $= 2.1$ μF

So charge flow in charging, $Q = CV = 2.1$ μF $\times$ 20 V $= 42$ μC

So the charges on the plates (from left to right) are: -42 μC, $+42$ μC, -42 μC, $+42$ μC

Q7 (a) Capacitance of series combination $= \dfrac{120 \times 40}{120 + 40} = 30$ μF

So total capacitance $= 30$ μF $+ 30$ μF $= 60$ μF

(b) (i) $Q = CV = 60$ μF $\times$ 12 V $= 720$ μC

(ii) **Either**: Capacitors in series in ratio 3 : 1
So pds in ratio 1 : 3, i.e. $\frac{1}{4}$ on the 120 μF
So the pd across the 120 μF $= 3$ V
Or: Charge on the series combination $= \frac{1}{2} \times 720$ μC $= 360$ μC
So charge on 120 μF $= 360$ μC

So pd on 120 μF $= \dfrac{Q}{C} = \dfrac{360\ \text{μC}}{120\ \text{μF}} = 3$ V

Q8 (a) The total separated charge on the capacitors is unchanged but it is now shared between them. Because the capacitors have equal value and the pds must be equal, they share the charge equally. So each had half the charge and the pd is half, i.e. 4.5 V.

(b) Initial energy on $C_1 = \frac{1}{2}CV^2 = 0.5 \times 50$ mF $\times (9.0\ \text{V})^2 = 2.025$ J [2.0 J to 2 sf]
Final energy $= 2 \times (0.5 \times 50\ \text{mF} \times (4.5\ \text{V})^2) = 1.0125$ J [1.0 J to 2 sf]
So energy change $= 1.0$ J $- 2.0$ J $= -1.0$ J [i.e. a 'loss' of 1.0 J]

Q9 (a) The final (maximum) charge. [CV_0 would be an acceptable answer.]

(b) (i) $Q = CV$. Substituting in $Q = Q_0(1 - e^{-t/RC})$ for Q and Q_0

∴ $CV_C = CV_0(1 - e^{-t/RC})$ and dividing by C gives the required equation.

(ii) $V_R = V_0 - V_C = V_0 - V_0\left(1 - e^{-t/RC}\right) = V_0 e^{-t/RC}$

(iii) $V_R = V_0 e^{-t/RC}$. Dividing by $R \longrightarrow \dfrac{V_R}{R} = \dfrac{V_0}{R} e^{-t/RC}$

$I = \dfrac{V_R}{R}$. When the capacitor is uncharged, $V_R = V_0$, so $I_0 = \dfrac{V_0}{R}$

Hence $I = I_0 e^{-t/RC}$

(c) (i) Capacitance, C, is defined by the equation $C = \dfrac{Q}{V}$, $[C]$ = A s V^{-1}

Resistance, R, is defined by the equation $R = \dfrac{V}{I}$, ∴ $[R]$ = V A^{-1}

∴ $[RC]$ = V A^{-1} × A s V^{-1} = s. ∴ $\left[\dfrac{Q_0}{RC}\right]$ = C s^{-1} = ampère

(ii) I_0 = gradient of tangent at $t = 0$.

∴ Gradient = $\dfrac{V_0}{R} = V_0 \times \dfrac{1}{R} = \dfrac{Q_0}{C} \times \dfrac{1}{R} = \dfrac{Q_0}{RC}$

Q10 (a)

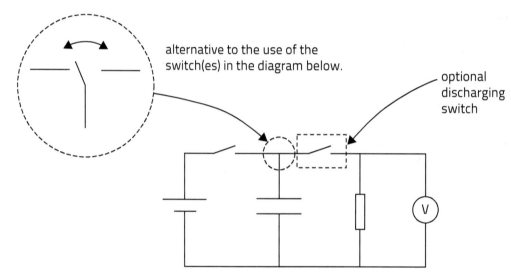

alternative to the use of the switch(es) in the diagram below.

optional discharging switch

(b) (i) Time for V_C to halve from 4.0 V to 2.0 V = 24.0 s – 8.5 s = 15.5 s
Time for V_C to halve from 2.0 V to 1.0 V = 39.5 s – 24.0 s = 15.5 s
Thus the half-life is constant, showing an exponential decay.

(ii) [Easy method]: After one time constant, the value of V_C falls to $1/e$ = 0.37 of the original value.
0.37 × 5.9 V = 2.2 V $\longrightarrow$ time = 22 s.

[More difficult, but equally valid]: $V_C = V_0 e^{-t/RC}$, so $\ln\left(\dfrac{V_0}{V_C}\right) = \dfrac{t}{RC}$

Choose, e.g. $t = 39$ s $\longrightarrow V_C = 1.0$ V, so $\ln\left(\dfrac{5.9}{1.0}\right) = \dfrac{39}{RC}$ $\longrightarrow RC$ (time constant) = 22 s

(c) $V_C = V_0 e^{-t/RC}$: taking logs ∴ $\ln (V_0/\text{volt}) = \ln (V_C/\text{volt}) - \dfrac{t}{RC}$

∴ Intercept = $\ln (V_C / \text{volt})$ = $\ln (5.9 \text{ V} / \text{V})$ = $\ln 5.9$ = 1.8 (2 sf)
and gradient = $-(RC)^{-1}$ = $-1/22$ = -0.045 (2 sf)

Section 4.2: Electric and gravitational fields of force

Q1

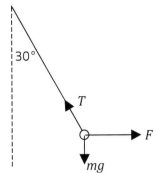

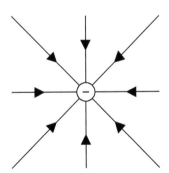

vector diagram

(a) From the vector diagram, $\tan 30° = \dfrac{F}{mg} = \dfrac{F}{2.00 \times 10^{-7} \times 9.81}$ ∴ $F = 1.13 \times 10^{-6}$ N $\sim 1.1 \times 10^{-6}$ N

(b) Separation of charges = 0.20 m using $F = \dfrac{1}{4\pi\varepsilon_0}\dfrac{Q_1Q_2}{r^2}$

$\dfrac{Q^2}{(0.2)^2}$

∴ $1.13 \times 10{-6} = 9.0 \times 10^9 \times$

∴ $Q = 2.2 \times 10{-9}$ C (2 sf)

Q2 (a) Assuming that the Moon is spherically symmetric, the field strength at its surface is given by:

$g = \dfrac{GM}{r^2}$, so $M = \dfrac{gr^2}{G} = \dfrac{1.62\,\text{N kg}^{-1} \times (1737 \times 10^3\,\text{m})^2}{6.67 \times 10^{-11}\,\text{N m}^2\,\text{kg}^{-2}} = 7.33 \times 10^{22}$ kg

(b) $F = \dfrac{GM_1M_2}{r^2} = \dfrac{6.67 \times 10^{-11} \times 5.97 \times 10^{24} \times 7.33 \times 10^{22}}{(3.84 \times 10^8)^2}$ N $= 1.98 \times 10^{20}$ N

Q3 (a) $F_G = \dfrac{GM^2}{r^2}$; $F_E = \dfrac{1}{4\pi\varepsilon_0}\dfrac{Q^2}{r^2}$,

so $\dfrac{F_E}{F_G} = \dfrac{1}{4\pi\varepsilon_0}\dfrac{Q^2}{GM^2} = 9.0 \times 10^9\dfrac{(1.60 \times 10^{-19})^2}{6.67 \times 10^{-11} \times (1.67 \times 10^{-27})^2} = 1.24 \times 10^{36}$ (3 sf)

(b) There are almost exactly the same number of protons and electrons on each of the Earth and the Sun, so these bodies are virtually electrically neutral. Hence the electrostatic force between them is negligible. The masses of the protons and electrons are both positive (there are no negative masses) so these both contribute positively to the gravitational force.

Q4 (a)

(b) P is equidistant from the two charges, so the magnitudes of the fields due to the charges are equal. The fields are directed radially towards the – charge and away from the + charge, so from the symmetry, the vertical components cancel, the horizontal components add and the resultant field is horizontal to the right.

Q5 (a) (i) $E = \dfrac{1}{4\pi\varepsilon_0}\dfrac{Q}{r^2} = 9.0 \times 10^9 \times \dfrac{(-3.0 \times 10^{-12})}{(0.12 \cos 35°)^2}$ South $= -2.79$ (N C^{-1}) or (V m^{-1})

i.e. 2.8 N C^{-1} (2 sf) due North.

(ii) $E = 9.0 \times 10^9 \times \dfrac{3.0 \times 10^{-12}}{(0.12 \sin 35°)^2}$ West = 5.7 N C^{-1} due West.

(iii) $E_R = \sqrt{(2.8)^2 + (5.7)^2} = 6.4 \text{ N C}^{-1}$

$\theta = \tan^{-1}\left(\dfrac{5.7}{2.8}\right) = 64°$. $\therefore$ Bearing $= 360° - 64°$

$\therefore 6.4 \text{ N C}^{-1}$ at $296°$

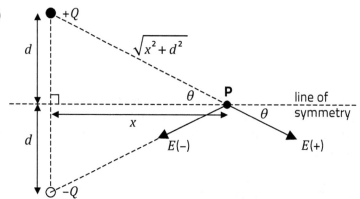

(b) (i) $V = \dfrac{1}{4\pi\varepsilon_0}\dfrac{Q}{r}$, $\therefore V_{tot} = 9.0 \times 10^9 \times 3.0 \times 10^{-12} \times \left(-\dfrac{1}{0.12 \cos 35°} + \dfrac{1}{0.12 \sin 35°}\right)$

$= 0.118 \text{ V}$

(ii) Initial potential energy of proton $= eV = 1.60 \times 10^{-19} \text{ C} \times 0.118 \text{ V} = 1.89 \times 10^{-20} \text{ J}$

$\therefore$ Max kinetic energy $= 1.89 \times 10^{-20} \text{ J}$ (when all PE has been lost)

$\therefore v = \sqrt{\dfrac{2E_k}{m_p}} = \sqrt{\dfrac{2 \times 1.89 \times 10^{-20} \text{ J}}{1.67 \times 10^{-27} \text{ kg}}} = 4760 \text{ m s}^{-1}$ (3 sf)

Q6 (a) **Either** The equipotential surfaces are at right angles to the field lines, which are radial.

Or The potential is given by: $V = \dfrac{1}{4\pi\varepsilon_0}\dfrac{Q}{r}$, so all points at the same radius, r, have the same potential.

(b) The potential is inversely proportional to the distance: $V = \dfrac{1}{4\pi\varepsilon_0}\dfrac{Q}{r}$, so $r \propto \dfrac{1}{V}$.

$\dfrac{1}{2} - \dfrac{1}{3} = \dfrac{1}{6}$ but $\dfrac{1}{3} - \dfrac{1}{4} = \dfrac{1}{12}$, which is only half as much.

Q7 (a) The defined zero of gravitational potential energy is infinity. Gravity is always an attractive force, hence work must be done to separate objects to infinity. Doing work on a system transfers energy to the system; hence for separations less than infinite (!) the potential energy (and therefore the potential at a non-infinite point) must be negative.
[Note: In an exam, the first two sentences in this answer would probably score you both marks (and there is not really enough room to write more) but the last sentence is needed for a full explanation.]

(b) (i) The equation $\Delta(\text{PE}) = mgh$ only applies in situations in which g may be taken as constant. In fact, g varies as r^{-2} so, for large increases in r, this must be taken into account.

(ii) Let mass of rocket be m. Then, applying conservation of energy:
Initial KE + Initial PE = PE at greatest height [because KE = 0 at greatest height].

$\dfrac{1}{2}m \times (3000)^2 - \dfrac{6.67 \times 10^{-11} \times 6.42 \times 10^{23}\,m}{3390 \times 10^3} = -\dfrac{6.67 \times 10^{-11} \times 6.42 \times 10^{23}\,m}{r}$

$\therefore -8.13 \times 10^6 = -\dfrac{4.28 \times 10^{13}}{r}$, $\therefore r = \dfrac{4.28 \times 10^{13}}{8.13 \times 10^6}$ m $= 5267$ km

$\therefore$ Height $= 5267 - 3390 = 1900$ km (2 sf)

Q8 (a) (i)

Magnitudes: $E(+) = E(-) = \dfrac{1}{4\pi\varepsilon_0}\dfrac{Q}{x^2 + d^2}$. The horizontal components of $E(+)$ and $E(-)$ are equal and opposite, to cancel. The vertical components add.

$\therefore E = \dfrac{1}{4\pi\varepsilon_0}\dfrac{2Q}{x^2 + d^2}\sin\theta = \dfrac{1}{4\pi\varepsilon_0}\dfrac{2Q}{x^2 + d^2}\dfrac{d}{\sqrt{x^2 + d^2}} = \dfrac{1}{2\pi\varepsilon_0}\dfrac{Qd}{(x^2 + d^2)^{3/2}}$

The direction of E is vertically downwards in the diagram.

(ii) The <u>resultant</u> electric field strength is always at right angles to the line of symmetry, so there is no component of the electric force on a test charge in the direction of motion (along the symmetry line) and so the work done is zero. Hence, Adam is right and Bethan is wrong.
Alternative answer: Work has to be done against the repulsion of $+Q$ in bringing a (positive) test charge from infinity. However, this is balanced by the equal negative work needed to bring the test charge against the attraction of $-Q$ and so the net work is zero. Adam is right and Bethan is wrong.

(b) For the $+/-$ charge combination, the electric field strength, E, is always at right angles to the line of symmetry; its maximum value is when $x = 0$ and it decreases towards zero as $x \longrightarrow \infty$.
For two equal positive charges the direction of E is always along the line of symmetry, in the $+x$ direction. E is zero when $x = 0$ (because the fields from the two charges are equal and opposite), rises to a maximum as x increases before decreasing towards zero as $x \longrightarrow \infty$ (inverse square at large distances).

Section 4.3: Orbits and the wider universe

Q1 Orbital period = 1 year = $60 \times 60 \times 24 \times 365.25$ s = 3.156×10^7 s

Orbital radius = 150×10^6 km = 1.50×10^{11} m

$$T = 2\pi \sqrt{\frac{r^3}{GM_\odot}}, \therefore M_\odot = \frac{4\pi^2}{T^2} \times \frac{r^3}{G} = \frac{4\pi^2}{(3.156 \times 10^7)^2} \times \frac{(1.50 \times 10^{11})^3}{6.67 \times 10^{-11}} = 2.0 \times 10^{30} \text{ kg}$$

Q2 (a) $g = \dfrac{GM}{r^2}, \therefore M = \dfrac{gr^2}{G} = \dfrac{9.81 \times (6.37 \times 10^6)^2}{6.67 \times 10^{-11}} = 5.97 \times 10^{24}$ kg

(b) The mass distribution is spherically symmetric.

(c) $\rho = \dfrac{M}{V} = \dfrac{5.97 \times 10^{24} \text{ kg}}{\frac{4}{3}\pi \times (6.37 \times 10^6 \text{ m})^3} = 5510 \text{ kg m}^{-3}$

Q3 The position of the star appears reasonable. It should be at one of the foci of the ellipse.
Given that S is correct, then X is clearly incorrect. It is the closest distance of approach between the planet and the star. Hence, this is the point at which the planet moves most quickly. It should be at the opposite end of the major axis [or S should be at the right-hand focus].

Q4 For both parts: 1 year = $60 \times 60 \times 24 \times 365.25$ s = $31\,557\,600$ s [$31\,536\,000$ if 365 days used]

(a) $T = 2\pi \sqrt{\dfrac{d^3}{GM}}$, so $d^3 = \dfrac{GMT^2}{4\pi^2} = \dfrac{6.67 \times 10^{-11} \times 1.99 \times 10^{30} \times (3.156 \times 10^7)^2}{4\pi^2}$

$\therefore d = 1$ AU = 1.496×10^{11} m = 150 million km (3 sf) [149 million km if 365 days used]

(b) Distance = $vt = 3.00 \times 10^5$ km s$^{-1} \times 3.156 \times 10^7$ s = 9.47×10^{12} km

[9.46×10^{12} km if 365 days used]

Q5 (a) Period = 27.3 day = $60 \times 60 \times 24 \times 27.3$ s = 2.36×10^6 s

$a = r\omega^2 = r\left(\dfrac{2\pi}{T}\right)^2 = 3.83 \times 10^8 \text{ m} \times \left(\dfrac{2\pi}{2.36 \times 10^6 \text{ s}}\right)^2 = 2.72 \times 10^{-3} \text{ m s}^{-2}$

(b) (i) ratio = $\dfrac{9.81}{2.72 \times 10^{-3}} = 3610$

(ii) ratio = $\left(\dfrac{3.83 \times 10^8 \text{ m}}{6.37 \times 10^6 \text{ m}}\right)^2 = 3620$

(c) At radius r_1: $g_1 = \dfrac{GM}{r_1^2}$. At r_2: $g_2 = \dfrac{GM}{r_2^2}$

$\therefore \dfrac{g_2}{g_1} = \left(\dfrac{r_1}{r_2}\right)^2$. As the data show, these ratios are equal to 2 sf. Hence they provide good support for Newton's law.

[The discrepancy can be attributed to use of data to 3 sf and to rounding.]

Q6 (a) $mr\left(\dfrac{2\pi}{T}\right)^2 = \dfrac{GMm}{r^2}$ or $T = 2\pi\sqrt{\dfrac{r^3}{GM}}$: 1 day = 60 × 60 × 24 s = 86 400 s

$\therefore r^3 = \dfrac{GMT^2}{4\pi^2} = \dfrac{6.67 \times 10^{-11} \times 5.97 \times 10^{24} \times 86\,400^2}{4\pi^2}$

$\therefore r = 42\,200\,000$ m

$\therefore$ Height above ground = 42 200 000 – 6 370 000 m = 35 800 km

(b) The orbit must be in the plane of the equator.

Q7 Kepler's 3rd law (for a circular orbit) states that radius^3 $\propto$ period^2

(using the given units) For Phobos: $\dfrac{\text{radius}^3}{\text{period}^2} = \dfrac{9.39^3}{0.319^2} = 8140$ (3 sf)

For Deimos: $\dfrac{\text{radius}^3}{\text{period}^2} = \dfrac{23.46^3}{1.263^2} = 8090$ (3 sf)

These values are very close. They are not the same to 3 sf but the 3 sf data are raised to higher powers, which amplifies errors and reduces the number of reliable sf. Hence K3 appears to be reasonably well followed.

Q8 (a) The density at which the universe will slow down asymptotically to zero at infinite expansion.

(b) $[G] = $ N m^2 kg^{-2} = kg m s^{-2} m^2 kg^{-2} = kg^{-1} m^3 s^{-2} ; $[H_0] = $ s^{-1}

$\therefore \left[\dfrac{3H_0^2}{8\pi G}\right] = \dfrac{\text{s}^{-2}}{\text{kg}^{-1}\,\text{m}^3\,\text{s}^{-2}} = $ kg m^{-3} = $[\rho]$, so dimensionally correct.

Q9 (a) $\dfrac{\Delta\lambda}{\lambda} = \dfrac{v}{c}$, so $v_{max} = \dfrac{3.00 \times 10^8 \times (393.82 - 393.36)}{393.36} = 3.51 \times 10^5$ m s^{-1} (350 km s^{-1})

$v_{min} = \dfrac{3.00 \times 10^8 \times (393.14 - 393.36)}{393.36} = -1.68 \times 10^5$ m s^{-1} (–170 km s^{-1})

(b) The star is orbiting a companion. The mean radial velocity is positive, so the system is moving away from the Earth. When the radial velocity is 350 km s^{-1} the star is on the part of its orbit where it is moving away from the Earth. When the radial velocity is –170 km s^{-1}, the star is on the part of its orbit where it is moving towards us, showing that the orbital speed is greater than the recession speed of the system.

Q10 (a) $T = 2\pi\sqrt{\dfrac{d^3}{GM}} = 2\pi\sqrt{\dfrac{(3.0 \times 10^{12})^3}{6.67 \times 10^{-11} \times 4.0 \times 10^{30}}} = 2.0(0) \times 10^9$ s

(b) Masses in ratio 3 : 5, so orbit radii in ratio 5 : 3

So Star 1 orbit has radius $\dfrac{5}{8} \times 3.0 \times 10^{12}$ m = 1.9 × 10^{12} m, and Star 2 orbit has radius 1.1 × 10^{12} m.

Q11 (a) Peak to peak range of velocities is +350 to –590 m s^{-1}: a range of 940 m s^{-1}

So orbital velocity = $\dfrac{1}{2} \times 940 = 470$ m s^{-1}.

(b) Period = 3.3 days = 285 000 s

$\therefore$ Circumference of orbit = 470 × 285 000 = 1.34 × 10^8 m

$\therefore$ Radius = $\dfrac{1.34 \times 10^8}{2\pi} = 2.13 \times 10^7$ m

(c) Period of planet's orbit = 285 000 s

$T = 2\pi\sqrt{\dfrac{r^3}{GM}}$, $\therefore r^3 = \dfrac{T^2 GM}{4\pi^2} = \dfrac{(2.85 \times 10^5)^2 \times 6.67 \times 10^{-11} \times 2.6 \times 10^{30}}{4\pi^2}$

$\therefore$ orbital radius, $r = 7.1 \times 10^9$ m

(d) Mass of planet = $\dfrac{2.13 \times 10^7}{(2.13 \times 10^7 + 7.1 \times 10^9)} \times 2.6 \times 10^{30}$ kg = 7.8 × 10^{27} kg

Q12 (a) Period of orbit, $T = 160$ day = 1.38 × 10^7 s

Speed of orbit, $v = 60$ km s^{-1}

$\therefore v = \dfrac{2\pi r}{T}$, $\therefore r_{vis} = \dfrac{vT}{2\pi} = \dfrac{60 \times 1.38 \times 10^7}{2\pi} = 1.32 \times 10^8$ km

(b) Centripetal force on BH = Gravitational force on BH

$$\therefore m_{BH}\left(\frac{2\pi}{T}\right)^2 r_{BH} = \frac{Gm_{BH}m_{vis}}{(r_{vis}+r_{BH})^2}$$

Dividing both sides by m_{BH} gives the required equation.

(c) With 8.4×10^{10} m:

$$\text{LHS} = \left(\frac{2\pi}{1.38 \times 10^7}\right)^2 \times 8.4 \times 10^{10} = 0.0174 \text{ m s}^{-1}$$

$$\text{RHS} = \frac{6.67 \times 10^{-11} \times 12 \times 10^{30}}{(2.2 \times 10^{11})^2} = 0.0165 \text{ m s}^{-1} \text{ which is the same as LHS to 2 sf.}$$

(d) Using 8.4×10^{10} m. $m_{BH} \times 8.4 \times 10^{10} = 12 \times 10^{30} \times 1.32 \times 10^{11}$

$$\therefore m_{BH} = 1.9 \times 10^{31} \text{ kg } [\sim 10 \text{ solar masses}]$$

Q13 (a) $H_0 = \text{gradient} = \dfrac{30\,000 \text{ km s}^{-1}}{480 \text{ Mpc}} = \dfrac{3.00 \times 10^7 \text{m s}^{-1}}{480 \times 3.09 \times 10^{22}\text{m}} = 2.02 \times 10^{-18} \text{ s}^{-1}$

(b) Motion of galaxies within clusters

Q14 If M is the total mass and d the separation: $T = 2\pi\sqrt{\dfrac{d^3}{GM}}$

$$\therefore M = \frac{4\pi^2 d^3}{GT^2} = \frac{4\pi^2(30 \times 1.50 \times 10^{11})^3}{6.67 \times 10^{-11} \times (82.2 \times 3.16 \times 10^7)^2} = 8.0 \times 10^{30} \text{ kg} = 4M_\odot$$

Radius of orbit of more massive star = $\frac{1}{4}$ of separation.

$\therefore$ Mass of less massive star = $\frac{1}{4} \times 4M_\odot = M_\odot$; mass of more massive star = $3M_\odot$

Section 4.4: Magnetic fields

Q1 (a) I = current; ℓ = length of wire; θ = angle between the wire and field

(b) $[F] = \text{kg m s}^{-2}$, $[I] = \text{A}$ and $[\ell] = \text{m}$. $\sin\theta$ has no unit.
$\therefore \text{T} = (\text{kg m s}^{-2}) \text{ A}^{-1} \text{ m}^{-1} = \text{kg s}^{-2} \text{ A}^{-1}$

(c) Fleming's left hand motor rule

Q2 (a)

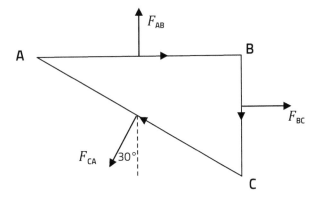

(b) (i) $\ell = 0.40 \cos 30° = 0.346$ m
$\therefore F_{AB} = BI\ell = 0.030 \times 2.5 \times 0.346 = 0.026$ N

(ii) $\ell = 0.40 \sin 30° = 0.20$ m, $\therefore F_{BC} = 0.015$ N

(iii) $F_{CA} = BI\ell = 0.030 \times 2.5 \times 0.40 = 0.030$ N

(c) Vertical component of $F_{AB} = 0.030 \times 2.5 \times 0.4 \cos 30° = 0.026$ N $= F_{AC}$
$\therefore$ Resultant vertical component = 0
Horizontal component of $F_{AB} = 0.030 \times 2.5 \times 0.4 \sin 30° = 0.015$ N $= F_{BC}$
$\therefore$ Resultant vertical component = 0

Q3 (a) Motor effect force: $F = BI\ell\sin\theta$, where θ = angle between wire and field.
AB is parallel to the field so $\sin\theta = 0$. Hence $F_{AB} = 0$
$BC = \ell\sin\phi$ and $\theta = 90°$, $\therefore F_{BC} = BI\ell\sin\phi$
For CA, $F_{CA} = BI\ell\sin\phi = F_{BC}$
By Fleming's LH motor rule F_{BC} is out of the paper and F_{AB} is into the paper, both at right angles. So forces are equal and opposite, so resultant force = 0 and Ella is correct.

(b) The lines of action of F_{AB} and F_{BC} are not the same so there is a resultant moment (i.e. a couple) and the triangle would rotate – BC out of the paper and AB into the paper. Hence Ella is correct (again!).

Q4 (a) (i) Note that B_Q is parallel to PR, B_R is parallel to PQ and B_P is parallel to QR

(ii) The resultant magnetic field is zero because the three fields are equal in magnitude and at 120° to one another.

The vector diagram is a closed triangle.

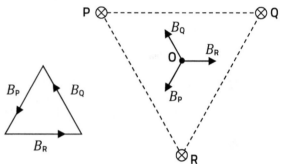

(b) (i) Field due to P at R = $\dfrac{\mu_0 I}{2\pi a} = \dfrac{4\pi \times 10^{-7} \times 7.5}{2\pi \times 0.060} = 2.5 \times 10^{-5}$ T.
$\therefore$ Force on 2.0 m of R = $BI\ell\sin\theta = 2.5 \times 10^{-5} \times 7.5 \times \sin 90° = 3.75 \times 10^{-4}$ N
Direction is towards P

(ii) Force due to Q = 3.75×10^{-4} N towards R.
$\therefore$ Resultant force = $2 \times 3.75 \times 10^{-4} \cos 30°$ N = 6.5×10^{-4} N vertically upwards.

Q5 (a) (i) The field is uniform and strongest in the central region of the solenoid and about half this strength at the ends of the solenoid.

(ii) Insert a Hall probe into the solenoid, orientated at right angles to the axis. Keeping the current constant, measure the Hall voltage (which is proportional to the magnetic field) at a series of positions along the length of the solenoid.

(b) B – right hand grip rule or corkscrew rule.

(c) $B = \mu_0 nI = \dfrac{4\pi \times 10^{-7} \times 300 \times 4.0}{0.60} = 2.5 \times 10^{-3}$ T

(d) By inserting an iron (or other ferromagnetic) core.

Q6 (a) By conservation of energy, $eV = \dfrac{1}{2}mv^2$,
$\therefore V = \dfrac{mv^2}{2e} = \dfrac{9.11 \times 10^{-31} \times (3.0 \times 10^7)^2}{2 \times 1.6 \times 10^{-19}} = 2560$ V

(b) $\dfrac{mv^2}{r} = Bev$, $\therefore B = \dfrac{mv}{er} = \dfrac{9.11 \times 10^{-31} \times 3.0 \times 10^7}{1.6 \times 10^{-19} \times 0.040} = 4.27 \times 10^{-3}$ T

Field at right angles into the diagram.

(c) If the forces balance, $Ee = Bev$, $\therefore E = 4.27 \times 10^{-3} \times 3.0 \times 10^7$ V m^{-1} = 128 kV m^{-1}
Direction = vertically downwards in the diagram.

Q7 (a) Force experienced = $Bqv\sin\theta$, at right angles to that of the velocity and field, in the direction given by the left-hand motor rule, where θ is the angle between the velocity and the field (here 90°) and v is the magnitude of the velocity.
As this force is always at right angles to the direction of motion, this provides the centripetal force.
Hence $\dfrac{mv^2}{r} = Bqv$, so $\dfrac{v}{r} = \dfrac{Bq}{m}$. But $\dfrac{v}{r} = \omega = \dfrac{2\pi}{T}$
$\therefore \dfrac{2\pi}{T} = \dfrac{Bq}{m}$ and hence $T = \dfrac{2\pi m}{Bq}$

(b) Frequency = $\dfrac{1}{\text{orbital period}} = \dfrac{0.30 \times 1.60 \times 10^{-19}}{2\pi \times 1.67 \times 10^{-27}} = 4.6 \times 10^6$ Hz

(c) The path of the protons [or other particles] is constant; the magnetic field is increased as the energy / momentum of the protons increases.
The synchrotron is used for producing very high energy (relativistic) particles, which themselves are used for probing the structure of subatomic particles.

Q8 (a) Total accelerating pd = 4×120 kV $= 4.8 \times 10^5$ V
∴ Increase in kinetic energy $= 4.8 \times 10^5$ eV $= 7.68 \times 10^{-14}$ J

$\frac{1}{2} \times 1.67 \times 10^{-27} \left(v^2 - (4.0 \times 10^6)^2\right) = 7.68 \times 10^{-14}$

∴ $v = 1.04 \times 10^7$ m s^{-1}

(b) (i) Each time a proton emerges from a tube, the next tube must be at a lower potential (so the proton can be accelerated). This can only happen if the pd reverses polarity each time the proton is within a tube.

(ii) As the proton gains energy it speeds up, so it travels further in the time taken for the polarity to reverse.

(c) It extends over a very long distance.

Section 4.5: Electromagnetic induction

Q1 (a) (i) $\Phi = \pi (0.04 \text{ m})^2 \times 0.050$ T $= 2.5 \times 10^{-4}$ Wb

(ii) $|\varepsilon| = \dfrac{\Delta \Phi}{\Delta t} = \dfrac{2.5 \times 10^{-4} \text{ Wb}}{0.16 \text{ s}} = 1.56$ mV $= 1.6$ mV (2 sf)

(iii) The emf is clockwise. We know this because a clockwise emf produces a clockwise current in the ring, and hence (using the right hand grip rule) a magnetic field inside the loop directed into the paper, opposing – as required by Lenz's law – the decrease in the applied field.

(iv) Energy = power × time$= \dfrac{\varepsilon^2}{R}\Delta t = \dfrac{(1.56 \times 10^{-3} \text{ V})^2}{2.75 \times 10^{-3} \,\Omega} \times 0.16$ s $= 1.4 \times 10^{-4}$ J

(b) Doubling the diameter, d, of the ring will quadruple its area (as $A = \pi \dfrac{d^2}{4}$). So the initial flux will quadruple, and so will the induced emf, ε. The energy dissipated is proportional to ε^2/R. But R is proportional to the diameter, so the energy dissipated is 8 times as much.

Q2 (a) (i) $|\varepsilon| = \dfrac{\Delta \phi}{\Delta t} = \dfrac{B\Delta A}{\Delta t} = \dfrac{Blv}{\Delta t} = Blv$

$= 0.35$ T $\times 0.15$ m $\times 0.20$ m s$^{-1} = 11$ mV

(ii) $I = \dfrac{0.0105 \text{ V}}{0.020 \,\Omega} = 0.53$ A (see diagram)

(iii) $F = BIl \sin \theta$
$= 0.35$ T $\times 0.525$ A $\times 0.15$ m $\times \sin 90°$
$= 27.5$ N $= 28$ mN (2 sf) (see diagram)

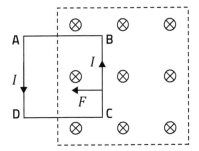

(b) Provided that the loop stays fully in the field, the flux linking it doesn't change as it moves, so there is no induced emf (Faraday's law).

(c) Left pointing arrow indicated <u>on side DA</u>.

Q3 (a) The direction of an emf induced by a change of flux is such that the effects of any current it produces oppose the change in flux.

(b) As we push the square loop into the field, there is an induced current, and energy is dissipated by resistive heating. But, as Lenz's law shows, the motor effect force on the loop is to the left, and whatever is pushing the loop has to expend energy doing work against this force. So energy is conserved.

(c) (i) Power dissipated, $\dfrac{\mathcal{E}^2}{R} = \dfrac{(0.0105\ \text{V})^2}{0.020\ \Omega} = 5.5\ \text{mW}$, using the emf calculated in 2(a)(i).

(ii) Work done per second = Fv = 0.0275 N × 0.20 m s^{-1} = 5.5 mW
using the force calculated in 2(a)(iii).

Q4 (a) (i) $\Phi = BA \cos \theta = 48 \times 10^{-6}\ \text{T} \times (0.12\ \text{m})^2 \times \cos 30° = 6.0 \times 10^{-7}\ \text{Wb}$

(ii) $N\Phi = 150 \times 6.0 \times 10^{-7}\ \text{Wb} = 9.0 \times 10^{-5}\ \text{Wb (turn)}$

(b) $|\mathcal{E}| = \dfrac{\Delta\Phi}{\Delta t} = \dfrac{(9.0 \times 10^{-5} - 0)\ \text{Wb-turn}}{1.2\ \text{s}} = 7.5 \times 10^{-5}\ \text{V}$

Q5 (a) When the magnetic flux linking a circuit changes an emf is induced in the circuit. The emf is proportional to the rate of change of flux linkage.

(b) (i) The flux linkage is proportional to cos θ. Therefore the rate of change of flux linkage, and hence (by Faraday's law), the emf, is proportional to sin θ. For example, when $\theta = \frac{\pi}{2}$ the flux linkage is zero, but its rate of change is a maximum, and so is the emf. The emf is also proportional to the area, PQ × QR, of the coil, the number of turns and the field strength (because these, as well as the angle, determine the flux linkage). By Faraday's law, the emf is also proportional to the rate at which the coil's angle changes, that is its angular velocity. The current is therefore proportional to all these factors, but is also inversely proportional to the sum of the resistances of the coil and resistor.

(ii) Note: from the information given, either graph is possible.

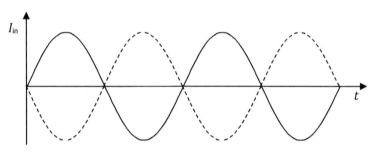

Q6 (a) As the magnet drops, the flux linked with the adjacent encircling wall of the tube changes, so an emf is induced in it, and a current, since the wall is conducting. The circular current makes a magnetic field inside the tube, roughly parallel to the coil axis. The magnet experiences a force from this field. According to Lenz's law, the force opposes the motion of the magnet.

(b) Because of the cut there can no longer be circular currents, so no magnetic field due to the induced emf and no magnetic force opposing the magnet's fall.

(c) Nabila is correct. The insulating glue will not interrupt the circular paths of currents due to the induced emfs, and there will still be a force opposing the magnet's motion, as explained in (a).

Q7 (a) Area swept out in time $\Delta t = l\ v\Delta t$ therefore in time Δt change in flux linking circuit = $\Delta\Phi = Bl\ v\Delta t$

$$\therefore \mathcal{E} = \frac{\Delta\Phi}{\Delta t} = \frac{Blv\Delta t}{\Delta t} = Blv$$

(b) (i) Initial emf = 0.25 T × 0.30 m × 0.50 m s^{-1}
= 0.0375 (37.5 mV) ∼ 40 mV

(ii) emf ∝ l, so increases linearly up to 75 mV. Hence (see graph):

(iii) length at 3 s = 0.525 m
∴ emf = 66 mV

$$\therefore I = \frac{\mathcal{E}}{R} = \frac{66\ \text{mV}}{1.50\ \Omega} = 44\ \text{mA}$$

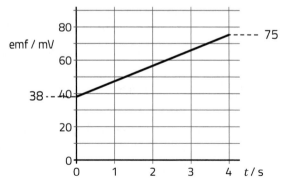

Option A: Alternating currents

Q1 (a) (i) 0.0000 s [or 0.0250 s, 0.0500 s, 0.0750 s.....]

(ii) $N\Phi = NBA$ = 50 turn × 0.150 T × $(0.040 \text{ m})^2$
= 0.0120 Wb turn

(b) (i) 0.0125 s [0.0375 s, 0.0625 s, 0.0875 s....]

(ii) $\mathcal{E}_{max} = BAN\omega$ = 0.0120 × 40.0π = 1.51 V [3 sf]

(c) (i) Time interval = 0.0020 s. Angle turned = $\frac{0.0020}{0.050}$ × 360° = 14.4°

∴ from -7.2° to 7.2°
Linkage at 0.0115 s = 0.012 sin (-7.2°) = -1.50 × 10^{-3} Wb turn
Linkage at 0.0135 s = 1.50 × 10^{-3} Wb turn
∴ $\Delta(N\Phi)$ = 3.00 × 10^{-3} Wb turn

(ii) $\langle\mathcal{E}\rangle = \frac{\Delta(N\Phi)}{t} = \frac{3.00 \times 10^{-3}}{0.0020}$ = 1.5 V

(iii) Almost the same, less than 1% difference. The rate of change of flux linkage is almost constant for small angles so to be expected.

Q2 (a) (i) $P = \frac{V_{rms}^2}{R}$, ∴ $V_{rms} = \sqrt{0.30 \times 5.6}$ = 1.30 V

(ii) pd across the internal resistance = $\frac{2.4}{5.6}$ × 1.30 V = 0.56 V
∴ emf = 1.30 V + 0.56 V = 1.86 V

(b) $\mathcal{E}_{rms} = \frac{BAN\omega}{\sqrt{2}} = \sqrt{2}\pi BANf$

∴ $f = \frac{1.86}{\sqrt{2}\pi \times 0.30 \times (0.05)^2 \times 120}$

= 4.7 Hz

Q3 (a) $V = \sqrt{V_R^2 + (V_C - V_L)^2}$

$= \sqrt{20^2 + (25 - 15)^2}$

= 22.4 V (3 sf)

(b) The reactance of the capacitor is more than that of the inductor. At resonance the two must be equal.

The reactance of the inductor goes up with frequency (down for the capacitor) so the resonance frequency is more than 500 Hz and Ciaran is correct.

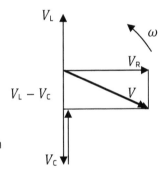

Q4 (a) Period = 4 × 2.0 ms = 8.0 ms ∴ Frequency = 125 Hz

(b) (i) Assuming uncertainty in reading y-axis = 0.1 div
Peak pd = (1.8 ± 0.2) div × 200 mV / div = 360 ± 40 mV
∴ rms pd = 250 ± 30 mV

(ii) Using 100 mV would double the amplitude of the trace which would halve the uncertainty because the uncertainty would be 0.1 in 3.6 rather than 0.1 in 1.8. So 100 mV / div would be better.

Q5 (a) Peak pds: V_R = 1.8 V; V_C = 3.0 V
∴ rms pds: V_R = 1.3 V; V_C = 2.1 V (2 sf)

(b) Power is only dissipated in the resistor.

$\langle P \rangle = \frac{I_0^2 R}{2} = \frac{(0.15 \text{ A})^2 \times 12 \, \Omega}{2}$ = 0.14 W (2 sf)

Q6 (a) [Peak values]: $V = I \times \dfrac{1}{2\pi fC}$ and $f = \dfrac{1}{T}$, so

$$I = \frac{2\pi CV}{T} = \frac{2\pi \times 0.60 \times 10^{-6} \times 10}{0.020} \text{ A} = 1.9 \text{ mA}$$

(b)
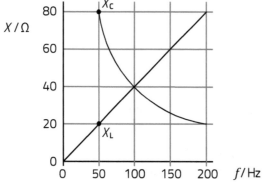

(c) From the definition of capacitance, $Q = CV$. The charging current, I, is the rate of change of Q, so is proportional to the rate of change of pd, V. Hence the maxima of the current occur at the times at which the rate of change of pd is greatest, i.e. 0 s, 0.02 s, and 0.04 s. [**Or**: the discharging current is greatest when the rate of change of pd is maximum negative, i.e. at 0.01 s and 0.03 s.]

Q7 RMS output pd, $V = \dfrac{BAN\omega}{\sqrt{2}}$

$$= \frac{1}{\sqrt{2}} \, 45 \times 10^{-3} \text{ T} \times 20 \times 10^{-4} \text{ m}^2 \times 240 \times \frac{1500 \times 2\pi}{60} \text{ s}^{-1}$$

$$= 2.40 \text{ V}$$

$\therefore$ Mean power, $\langle P \rangle = \dfrac{V_{\text{rms}}^2}{R} = \dfrac{(2.40 \text{ V})^2}{120 \ \Omega} = 0.048 \text{ W}$

Q8 (a) Both have the unit ohm or both equal to $\dfrac{V_{\text{rms}}}{I_{\text{rms}}}$.

(b) Reactance (X) is frequency-dependent; resistance (R) is not or $R = \dfrac{V(t)}{I(t)}$ but usually $X \neq \dfrac{V(t)}{I(t)}$

Q9 (a)
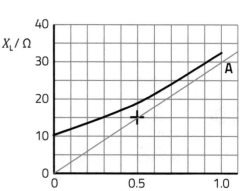

(b) (i) $X_C = \dfrac{1}{2\pi fC}$, so $C = \dfrac{1}{2\pi fX_C} = \dfrac{1}{2\pi \times 50 \times 80} = 4.0 \times 10^{-5}$ F

(ii) $X_L = 2\pi fL$, so $L = \dfrac{X_L}{2\pi f} = \dfrac{20}{2\pi \times 50} = 0.064$ H

Q10 (a) Graph **A** is the variation of the reactance, X_L, of the inductor with frequency.

(b) At $f = 0$, $Z = 10 \ \Omega$.
At $f = 0.5$ kHz, $Z = \sqrt{10^2 + 15^2} = 18.0 \ \Omega$
At $f = 1.0$ kHz, $Z = \sqrt{10^2 + 30^2} = 31.6 \ \Omega$

Q11 (a) (i)

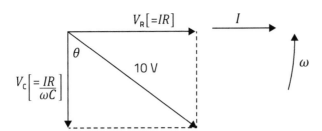

(ii) Applying Pythagoras' theorem: $10\ V = \sqrt{V_R{}^2 + V_C{}^2} = I\sqrt{R^2 + X_C{}^2}$

$$X_C = \frac{1}{2\pi \times 750 \times 0.47 \times 10^{-6}}\ \Omega = 450\ \Omega$$

$$\therefore I = \frac{10.0}{\sqrt{330^2 + 450^2}}\ A = 0.018\ A \sim 20\ mA$$

(b) (i) $V_C = IX_C = 0.018 \times 450 = 8.1\ V$

(ii) Angle θ in phasor diagram $= \tan^{-1}\left(\dfrac{V_R}{V_C}\right) = \tan^{-1}\left(\dfrac{R}{X_C}\right) = \tan^{-1}\left(\dfrac{330}{450}\right) = 36°$

(iii) This is not true: $V_R = \sqrt{10.0^2 - V_C{}^2} = 5.9\ V$ which is not the same as $(10.0 - 8.1)\ V$.

(c) (i) Only the resistor dissipates energy.
Power dissipated $= I^2R = 0.105\ W$

$\therefore$ Energy over one cycle $= \dfrac{0.105\ W}{750\ Hz} = 1.4 \times 10^{-4}\ J$ (2 sf)

(ii) Peak pd across capacitor $= 11.5\ V$
$\therefore$ Maximum energy stored $= \frac{1}{2}CV^2 = 3.1 \times 10^{-5}\ J$
$\therefore$ Mean energy $= 1.5 \times 10^{-5}\ J$

Q12 (a) Set the signal generator to 100 Hz. Take simultaneous readings of the current and pd using the ammeter and voltmeter respectively. Divide the pd by the current to give the impedance.

(b)

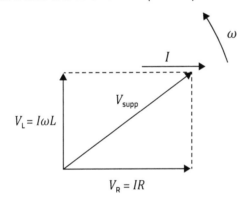

From the phasor diagram: $V_{supp} = \sqrt{V_L{}^2 + V_R{}^2} = I\sqrt{X_L{}^2 + R^2}$
By definition $V_{supp} = IZ \therefore Z^2 = X_L{}^2 + R^2 = (2\pi f L)^2 + R^2$

(c) (i) Intercept $= 25\ \Omega^2$
$\therefore R = 5.0\ \Omega$

(ii) Gradient $= \dfrac{231 - 25}{10 \times 10^5} = 2.06 \times 10^{-4}\ \Omega^2\ s^2$

Gradient $= 4\pi^2 L^2, \therefore L^2 = 5.22 \times 10^{-6}\ H^2$
$\therefore L = 2.3\ mH$

Q13 (a) At resonance, pd across $R = 5.00\ V$

$\therefore R = \dfrac{5.00}{1.00}\ \Omega = 5.00\ \Omega$

Resonance frequency $f_{res} = \dfrac{1}{2\pi\sqrt{LC}} \therefore L = \dfrac{1}{4\pi^2 f_{res}{}^2 C} = 2.25 \times 10^{-3}\ H$

(b) (i) $V_C = \dfrac{I}{2\pi f C} = \dfrac{1.00}{2\pi \times 10.6 \times 10^3 \times 0.100 \times 10^{-6}} = 150\ V$

(ii) $Q = \dfrac{150}{5} = 30$

Q14
$$L = 4\pi \times 10^{-7} \text{ Hm}^{-1} \times \frac{25^2 \times \pi \times (3.0 \times 10^{-3} \text{ m})^2}{15 \times 10^{-3} \text{ m}} = 1.48 \times 10^{-6} \text{ H}$$

$$f_{res} = \frac{1}{2\pi\sqrt{LC}}, \text{ so } C = \frac{1}{4\pi^2 f_{res}^2 L} = \frac{1}{4\pi^2 + (1.6 + 10^6 \text{ Hz})^2 + 1.48 + 10^{-6}\text{H}} = 6.7 \text{ nF}$$

Q15 (a)

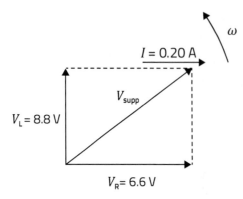

$$V_R = 0.20 \text{ A} \times 33 \text{ }\Omega = 6.6 \text{ V}$$
$$X_L = 2\pi \times 100 \text{ Hz} \times 0.070 \text{ H} = 44.0 \text{ }\Omega$$
$$\therefore V_L = 0.20 \text{ A} \times 44 \text{ W} = 8.8 \text{ V}$$
$$\therefore V_{supp} = \sqrt{6.6^2 + 8.8^2} = 11 \text{ V}$$

(b) (i) $f_{res} = \dfrac{1}{2\pi\sqrt{LC}}$

$$\therefore C = \frac{1}{4\pi^2 f_{res}^2 L} = \frac{1}{4\pi^2 \times (100 \text{ Hz})^2 \times 0.070\text{H}}$$

$$= 3.6 \times 10^{-5} \text{ F} = 36 \text{ }\mu\text{F}$$

(ii) $I = \dfrac{V}{R} = 0.30 \text{ A}$

(c) He will be correct as far as the peak current is concerned because both the pd and resistance are doubled. However the resonance curve will be less sharp (the Q factor will be less) because the ratio of the pds across the inductor (or capacitor) and resistor will be less.

Option B: Medical physics

Q1 X-ray attenuation in tissues increases with increasing density, so bones (high density) cause more attenuation than soft tissues, making good contrast images. X-rays have very low wavelength so diffraction is negligible, so produce sharp images. [They can also be detected by CCD devices, photographic film or MOSFET devices.]

Q2 (a) The electrons are accelerated to a high energy by the pd. These high energy (70 keV) electrons decelerate rapidly in the tungsten target. Charged particles which accelerate produce radiation [by the process of *bremsstrahlung*] giving the continuous spectrum.

(b) The electrons sometimes knock out an inner electron from a tungsten atom. Higher energy electrons within the atoms drop down into the vacant energy level giving out photons with energy equal to the difference in the two energy levels.

(c) [A vacuum is needed] to allow the electrons to pass down the tube without colliding with air molecules and losing energy.

(d) $\frac{1}{2}mv^2 = eV$, so $v = \sqrt{\dfrac{2eV}{m}} = \sqrt{\dfrac{2 \times 1.60 \times 10^{-19} \text{ C} \times 70 \times 10^3 \text{ V}}{9.11 \times 10^{-31}\text{kg}}} = 1.6 \times 10^8 \text{ m s}^{-1}$

This calculated speed is too close to the speed of light (in a vacuum) for the above equations to be valid.

(e) $eV = \dfrac{hc}{\lambda_{min}}$, so $\lambda_{min} = \dfrac{hc}{eV} = \dfrac{6.63 \times 10^{-34} \text{ Js} \times 3.00 \times 10^8 \text{ m s}^{1}}{1.60 \times 10^{-19} \text{ C} \times 70 \times 10^3 \text{ V}} = 1.8 \times 10^{-11} \text{ m}$

(f) (i) Power input = 14.5 mA × 70 kV = 1015 W
   ∴ Efficiency = $\dfrac{5.1\ \text{W}}{1015\ \text{W}}$ = 5.0 × 10^{-3} [= 0.5%]

(ii) Heat needs to be conducted away at the rate of 1015 W - 5.1 W, i.e. about 1 kW. In the absence of cooling water the temperature of the tungsten cathode would become too high and the X-ray tube would be destroyed (the glass would melt).

(iii) Predicted % efficiency = 70 × 74 × 10^{-4} = 0.52 (2 sf)
So this is a very good approximation.

Q3

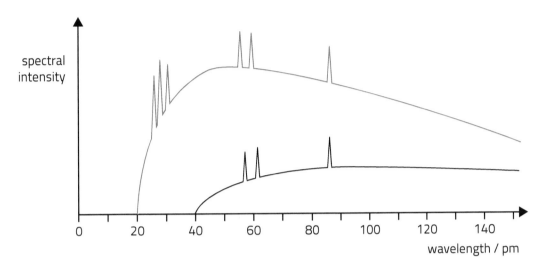

Q4 (a) (i) In the piezoelectric effect, distorting a crystal produces a pd. If a pd is applied to a piezoelectric crystal, it produces a distortion – the reverse piezoelectric effect.

(ii) A short-duration high frequency (MHz) voltage applied to the crystal in the transducer causes rapid vibrations (of the same frequency) producing the ultrasound wave.

(b) A-scans are amplitude scans. The returning wave is displayed on a CRO and the time delay indicates the depth of the structure.
B-scans are brightness scans. The returning waves build up an image of the structures on a screen. [This is achieved using a large array of ultrasound detectors. 2D images are built up as the emitter and detectors are rotated/scanned.]

(c) (i)

Body tissue	Density / kg m^{-3}	Speed of sound / m s^{-1}	Acoustic impedance / kg m^{-2} s^{-1}
Soft tissue	1070	1590	1.70 × 10^6
Bone	1650	4080	6.73 × 10^6

(ii) $R = \dfrac{(6.73 - 1.70)^2}{(6.73 + 1.70)^2}$ =0.356 = 36% (2 sf)
[Note – the factors of 10^6 cancel so there is no need to include them in the calculation.]

(iii) There are four transitions between soft tissue and bone: skin/bone; bone/brain; brain/ bone; bone/skin. The fraction transmitted at each boundary = 0.644, so for four boundaries the transmission is (0.644)^4 = 0.172 = 17%. Hence the signal is not 13% of the baby's signal. So, the final figure is approximately correct, but this is a fluke as it does not arise from 1.00 − 0.36^2.

Q5 (a) μ is defined by: $I = I_o e^{-\mu x}$ where I is the intensity.
If $I = \frac{1}{2} I_o$, then $e^{-\mu x_{1/2}} = \frac{1}{2}$
Taking natural logs $\rightarrow -\mu x_{1/2} = \ln\frac{1}{2} = -\ln 2$
∴ $\mu x_{1/2} = \ln 2$

(b) $\mu = \dfrac{\ln 2}{3.7\ \text{cm}}$ = 0.1873 cm^{-1}

∴ $\dfrac{I}{I_0} = e^{-0.1873 \times 5.0}$ = 0.39

∴ Fractional reduction = 1.00 − 0.39 = 0.61 = 61%

Q6 (a) The scanner head produces pulses of ultrasound which penetrate the body and are reflected from the foetus. The reflected waves are detected by the scanner head, which gives out voltage signals to an analyser which builds up an image as the scanner is moved across the abdomen. The gel is required so that the reflection coefficient between the head and the abdomen is low – otherwise the air gap would produce large reflections and very little signal would get through to the foetus.

(b) (i) The ultrasound rays are reflected from the red blood cells back to the scanner head. If the blood cells are moving towards the scanner, the waves are Doppler shifted to a higher frequency: the blood cells 'see' waves of a higher frequency which they reflect; as the cells are a moving source the waves received by the scanner are further Doppler shifted. The frequency change is proportional to the speed of the blood cells.

(ii) $\frac{\Delta f}{f_0} = -\frac{2 \times 1.058 \text{ m s}^{-1}}{1580 \text{ m s}^{-1}} \cos 5° = 1.334 \times 10^{-3}$
$\therefore \Delta f = 1.334 \times 10^{-3} \times 5.50 \text{ MHz} = 7.34 \text{ kHz (3 sf)}$

Q7 (a) The anti-scatter grid is to cut out X-rays which are diverted at an angle by the material of the body – which interfere with the image.
The scintillator screen flashes at a point where an X-ray photon hits to produce the image.

(b) They are designed so that each X-ray photon produces a large number of visible photons, which reduces the dose of X-rays required. The screen is placed in a dark container so that even a very faint image will be picked up by the camera – also cutting down the X-ray dose required.

Q8 (a) In therapy, the photons are required to deliver a large energy to the target cells. Higher energy X-ray photons have a greater penetration. This means that the intensity of the beam drops more slowly, giving a more uniform dosage to the desired area.

(b) In therapy, the X-rays are required to cause damage to the target (tumour) cells. Higher intensity beams deliver a greater number of photons per second, increasing the damage to the target cells.

Q9 (a) **Photon explanation:**
The hydrogen nuclei have two spins (up and down in the magnetic field) with an energy difference, and more nuclei in the lower energy level. The radio photons have energy equal to this difference and promote a higher number of nuclei into the upper energy level. When the radio beam is turned off, the subsequent release of energy as the nuclei return to the lower energy level is detected.
Classical explanation:
The spinning hydrogen nuclei precess at a frequency proportional to the magnetic field strength. If radio waves with this precession frequency are incident they are strongly absorbed and re-emitted when the external radio waves are turned off.

(b) $f = 62.4 \, B$
$f_1 = 62.4 \times 1.53 = 95.5 \text{ MHz}; \, f_2 = 62.4 \times 1.92 \text{ MHz} = 120 \text{ MHz}$

(c) **MRI scans** produce high resolution (around 1 mm) 3D images in which the different tissues in the joint are distinguished. They require the use of very expensive equipment and can cause claustrophobia. They carry no known risk but cannot be given to people with metal implants, e.g. heart pacemakers.
Traditional X-rays only image bones clearly, with excellent image resolution (~0.1 mm) and are 2D. So the soft tissues (cartilage and ligaments) in the joint are not imaged well. They also have a small risk because of the ionising radiation involved. [3D and soft tissue images can be produced by CT X-ray scans with the dangers of higher radiation dose. Image resolution is around 0.5 mm.]
Ultrasound B-scans are cheap and benign but have poor resolution (usually around 2-5mm) and can image the soft tissues and the surface of bones (so are good for checking cartilage and ligament).

Q10 (a) Additional distance from **X** to detector **B** = 3.00×10^8 m s^{-1} × 237 ps = 7.11 cm
$\therefore$ **X** is 3.6 cm closer to **A** than the halfway point.

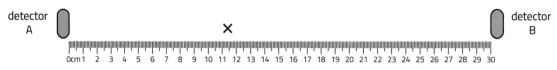

(b) A positron emitter is attached to glucose molecules where it is preferentially absorbed by actively dividing cells, such as tumour cells. The emitted positrons have short range and mutually annihilate with an electron in the body to produce 2 γ photons which emerge in opposite directions. The time difference between the detection of these photons enables the position of the emission site to be determined. A (3D) image is gradually built up showing the location of hot-spots where the annihilations take place.

Q11 The X-ray-emitting head and the detector rotate about the body and gradually move along the axis of the body. In this way the scanner produces images of the body in slices which can be combined in a computer to produce 3D images. Traditional X-rays don't distinguish soft tissues well but contrast agents can be injected into blood vessels or swallowed into the alimentary canal to improve the visibility of different soft tissues.

Q12 (a) (i) Effective dose = equivalent dose × tissue weighting factor
Effective dose (contribution for the liver) = 550 mSv × 0.04 = 22 mSv

(ii) Equivalent dose = absorbed dose × radiation weighting factor
Weighting factor for these neutrons = 20

Absorbed dose = $\dfrac{550}{20}$ = 27.5 mGy

Absorbed dose = $\dfrac{\text{energy of absorbed radiation}}{\text{mass}}$

∴ Energy of absorbed radiation = absorbed dose × mass
= 27.5 mGy × 94 kg = 2.6 J

(b) The radiation weighting factor of a radiation is 20, which is the highest, reflecting the fact that it is strongly ionising and hence it deposits its energy in a very short distance. The ingested material will come into close contact with the oesophagus, the stomach and the colon (even assuming none is absorbed into the bloodstream) which presents a high risk of cancers to the lining of these organs. Together these organs make up about 30% of the sensitivity of the body to radiation.

Q13 (a) The collimator is there to restrict the γ rays to those in the vertical direction, to allow an image to be made (because γ rays cannot be focused).

The scintillator produces a flash with many visible photons when one γ photon hits it allowing an image to be constructed. The photomultiplier is very sensitive and ensures that each of these flashes is detected to allow a low dose of ^{99m}Tc to be used.

(b) It needs to be non-toxic, have a short half-life and to produce radiation which is not absorbed in the body, which can hence emerge to be detected, i.e. a gamma emitter.

Option C: The physics of sports

Q1 Football is a contact sport in which players receive forces from the side. The shorter the height, h, (see diagram) the smaller the moment of F and the wider the base w, the bigger the opposing moment to being toppled. Hence the greater the stability.

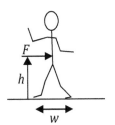

Q2 Clockwise moment of leg + foot = 118 × 43 + 11 × 91 N cm
= 6075 N cm

∴ By, principle of moments, F × 4.5 cm = 6075 N cm

∴ $F = \dfrac{6075\ \text{N cm}}{4.5\ \text{cm}}$ = 1350 N

Q3 (a) (i) Total $E_k = \sum \frac{1}{2}(\Delta m)v^2$, where the Δm are the small masses that make up the cylinder.

Every part of the cylinder is moving with the same speed, $v = rw$.

∴ $E_k = \frac{1}{2}v^2 \sum \Delta m = \frac{1}{2}(r\omega)^2 \sum \Delta m$

But $\sum \Delta m$ = total mass, m. ∴ $E_k = \frac{1}{2}mr^2\omega^2$

(ii) For a rolling cylinder moving with speed v, the angular velocity, $\omega = \frac{v}{r}$, so that the point of contact is stationary.

kinetic energy $E_k = \frac{1}{2}mv^2 + \frac{1}{2}I\omega^2$

But $I = \frac{1}{2}mr^2 \therefore \frac{1}{2}I\omega^2 = \frac{1}{2}mr^2\omega^2 = \frac{1}{2}m(r\omega)^2 = \frac{1}{2}mv^2$

Hence the rotational and linear KEs are equal, i.e. they contribute half each.

(iii) Loss of gravitational potential energy = $mg\Delta h$

Total KE = 2×0.45 J = 0.90 J

$\therefore$ By conservation of energy, $m = \dfrac{0.90\text{ J}}{9.81\text{ N kg}^{-1} \times 0.30\text{ m}} = 0.31$ kg (2 sf)

(b) For a particular mass and radius, the moment of inertia of the snooker ball is less than that of the cylinder, so the rotational kinetic energy of a rolling snooker ball is less than its translational kinetic energy. So for a particular potential energy loss the translational kinetic energy of the snooker ball is greater than that of the cylinder, so it has a greater speed and hence it arrives first.

Q4 (a) Linear acceleration, $a = \dfrac{[48\text{ m s}^{-1} - (-32\text{m s}^{-1})]}{0.0068\text{ s}} = 11\,800$ m s^{-2} Due west

$\Delta\omega = \dfrac{2550 - 1200}{60}$ rad s^{-1} = 22.5 rad s^{-1}

$\therefore$ Angular acceleration, $a = \dfrac{22.5\text{ rad s}^{-1}}{0.0068\text{ s}} = 3310$ rad s^{-2}

(b) Mean force, $\langle F \rangle = ma = 0.0582$ kg $\times$ 11 800 m s^{-2} = 687 N

Mean torque, $\langle \tau \rangle = I\alpha = \frac{2}{3} \times 0.0582$ kg $\times \left(\dfrac{66.9 \times 10^{-3}\text{ m}}{2}\right)^2 \times 3310$ rad s^{-2} = 0.144 N m

(c) Rotational kinetic energy,

$E_{k\text{ rot}} = \frac{1}{2}I\omega^2 = \frac{1}{2} \times \frac{2}{3} \times 0.0582$ kg $\times \left(\dfrac{66.9 \times 10^{-3}\text{ m}}{2}\right)^2 \times \left(\dfrac{2550\text{ rad}}{60\text{ s}}\right)^2 = 0.039$ J

Translational kinetic energy,

$E_{k\text{ trans}} = \frac{1}{2}mv^2 = \frac{1}{2} \times 0.0582$ kg $\times (48\text{ m s}^{-1})^2 = 67$ J

So Charles is correct by a large margin!

(d) Consider the vertical motion with upwards positive

To calculate the time to hit the ground: $y = u_yt - \frac{1}{2}gt^2$, with $y = -0.95$ m

$\therefore 4.905t^2 - 48t\sin 6.5° - 0.95 = 0$, $\therefore t = \dfrac{5.43 \pm \sqrt{29.5 + 4 \times 4.905 \times 0.95}}{9.81} = 1.26$ s

[ignoring the negative root]. $\therefore$ Range = (48 cos 6.5° $\times$ 1.26) m = 60 m (2 sf)

So she has hit it too far.

(e) (i) $F_D = \frac{1}{2}\rho v^2 A C_D$

$= \frac{1}{2} \times 1.25$ kg m$^{-3} \times (48\text{ m s}^{-1})^2 \times \pi\left(\dfrac{66.9 \times 10^{-3}\text{ m}}{2}\right)^2 \times 0.60 = 3.0$ N (2 sf)

Acting over a distance of 10 m this would reduce the KE by 30 J which is nearly half so it cannot be ignored.

(ii) Gravitational force = mg = 0.57 N, so the 'lift' is almost 4$\times$ as great and the range will be a lot less than the calculated 60 m.

Q5 The shower spray makes the air within the shower cubicle move. This causes the pressure within the cubicle to drop according to the Bernoulli equation, $p = p_0 - \frac{1}{2}\rho v^2$. The pressure difference between the inside and outside produces a net inward force on the shower curtain.

Q6 The anticlockwise moment of the weight of the wind surfer (the person and the vessel) about the contact with the water (the turning point, **T**) is balanced by the clockwise moment of the force of the wind on the sail.

force due to wind

weight ▼   T

Q7 Speed of approach = 11.8 m s^{-1}
Speed of separation = (11.4 – 0.4) m s^{-1} = 11.0 m s^{-1}
$\therefore$ Coefficient of restitution, $e = \dfrac{11.0 \text{ m s}^{-1}}{11.8 \text{ m s}^{-1}} = 0.93$

Q8 In the absence of external torques, the angular momentum, L, of the gymnast is constant.
$L = I\omega$, where $I = \sum mr^2$ is her moment of inertia.
In the tuck position, the distance r from the centre to many points of the body is reduced. Hence, I is reduced and ω, the angular velocity, increases.

Q9 (a) Momentum change,
$\Delta p = 0.058 \text{ kg} \times [49 \text{ m s}^{-1} - (-63 \text{ m s}^{-1})] \text{ m s}^{-1} = 6.50 \text{ N s}$ due south.
$\therefore$ Mean force on ball, $\langle F \rangle = \dfrac{\Delta p}{t} = \dfrac{6.5 \text{ N s}}{6.5 \times 10^{-3} \text{ s}} = 1.0 \times 10^3 \text{ N}$

(b) (i) $e = \dfrac{49 \text{ m s}^{-1}}{63 \text{ m s}^{-1}} = 0.78$

(ii) $e = \sqrt{\dfrac{\text{bounce height}}{\text{drop height}}}$, $\therefore \dfrac{\text{bounce height}}{\text{drop height}} = 0.78^2 = 0.60$

Q10 (a) $\langle F \rangle = \dfrac{m\Delta v}{t} = \dfrac{0.04593 \text{ kg} \times 85 \text{ m s}^{-1}}{257 \times 10^{-6} \text{ s}} = 15\,200 \text{ N}$

(b) $\Delta\omega = \dfrac{2700}{60} \times 2\pi \text{ rad s}^{-1} = 283 \text{ rad s}^{-1}$

$\therefore \langle \alpha \rangle = \dfrac{\Delta\omega}{t} = \dfrac{283}{257 \times 10^{-6}} \text{ rad s}^{-2} = 1.10 \times 10^6 \text{ rad s}^{-2}$

(c) Mean torque, $\langle \tau \rangle = I\alpha$
$\therefore \langle F_{\text{tan}} \rangle r = \tfrac{2}{5}mr^2 \langle \alpha \rangle$,
so $\langle F_{\text{tan}} \rangle = \dfrac{2mr}{5}\langle \alpha \rangle = \dfrac{2 \times 0.04593 \text{ kg} \times 21.34 \times 10^{-3}\text{m}}{5} \times 1.10 \times 10^6 \text{ rads}^{-2}$
$= 431 \text{ N}$

(d)

golf club head

F

F_{tan}

(e) $E_{\text{k rot}} = \tfrac{1}{2}I\omega^2 = \tfrac{1}{2} \times \tfrac{2}{5} \times 0.04593 \text{ kg} \times \left(21.34 \times 10^{-3} \text{ m}\right)^2 \times \left(283 \text{ rad s}^{-1}\right)^2 = 0.34 \text{ J}$

$E_{\text{k trans}} = \tfrac{1}{2}mv^2 = \tfrac{1}{2} \times 0.04593 \text{ kg} \times \left(85 \text{ m s}^{-1}\right)^2 = 166 \text{ J}$

$\therefore \dfrac{E_{\text{k rot}}}{E_{\text{k trans}}} = \dfrac{0.34 \text{ J}}{166 \text{ J}} = 2.0 \times 10^{-3}$

Q11 (a) Initial vertical velocity = 18.1 sin 21.2$°$ m s^{-1} = 6.545 m s^{-1}

Calculate time in air: $t = \dfrac{v-u}{a} = \dfrac{6.545 \text{ m s}^{-1} - (-6.545 \text{ m s}^{-1})}{9.81 \text{ m s}^{-1}} = 1.334 \text{ s}$

Horizontal velocity = 18.1 cos 21.2$°$ = 16.88 m s^{-1}
$\therefore$ Range = 22.5 m

(b) $F_{\text{D}} = \tfrac{1}{2}\rho v^2 A C_{\text{D}}$
$= \tfrac{1}{2} \times 1.25 \text{ kg m}^{-3} \times (18.1 \text{m s}^{-1})^2 \times \pi\,(0.110 \text{ m})^2 \times 0.195 = 1.52 \text{ N}$

(c) Because of the spin, the air passing round the ball is deflected downwards as shown in the diagram, acquiring a downward momentum. So the ball exerts a downward force on the air (N2) and the air exerts an equal upward force on the ball (N3) so the flight time of the ball is greater.

(d)

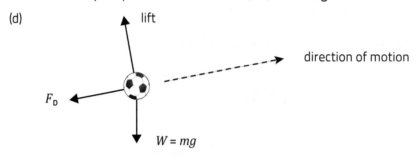

(e) The drag acts in the opposite direction to the motion of the ball through the air. In the absence of spin it produces a deceleration both horizontally and vertically. The vertical deceleration has a very small effect on the flight time because, although the mean upward velocity is less, the mean downward velocity is reduced further. The range, however, is reduced because of the horizontal deceleration producing a lower mean speed.

Backspin produces a lift force which is mainly vertical upwards, this increases the time the ball spends in the air and hence increases the horizontal range.

(f) Angular momentum is conserved if there is no externally applied torque [couple]. In this case, there is a couple applied to the ball by the air, so the principle does not apply to the ball on its own.

(g)

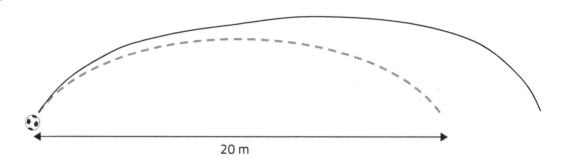

20 m

Option D: Energy and the environment

Q1 (a) (i) Cross-sectional area of asteroid = $\pi r^2 = \pi \times (150\,m)^2$
∴ Radiation absorption rate = $0.9 \times \pi \times (150\,m)^2 \times 1361\,W\,m^{-2}$
= 86.6 MW

(ii) If the asteroid is rotating, the surface temperature will be approximately the same on both faces. If it radiates as a black body, Stefan's law is valid.
i.e. $P = A\sigma T^4$, with $A = 4\pi r^2$
∴ $T^4 = \dfrac{86.6 \times 10^6\,W}{4\pi \times (150\,m)^2 \times 5.67 \times 10^{-8}\,W\,m^{-2}\,K^{-4}}$
∴ $T = 271\,K$

Alternatively:
$0.9 \times \pi \times 150^2 \times 1361 = 4\pi \times 150^2 \times 5.67 \times 10^{-8}\,T^4$
leading to the same answer

(iii) $\lambda_{peak} = \dfrac{W}{T} = \dfrac{2.90 \times 10^{-3}\,m\,K}{271\,K} = 10.7\,\mu m$
Infra-red.

(b) (i) The radiation from the surface passes through the atmosphere. This radiation is in the thermal infra-red and is partly absorbed by greenhouse gases in the atmosphere. This absorbed energy is re-radiated, in all directions, with 50% downwards where it is absorbed by the Earth, resulting in a higher stable temperature.

(ii) Higher concentrations result in a greater fraction of the radiation from the ground being absorbed and re-radiated downwards. Thus the ground temperature rises.

Q2 (a) A body which is wholly or partially immersed in a fluid experiences an upthrust equal to the weight of fluid displaced by the body.

(b) (i) Mass of ice $= 920 \text{ kg m}^{-3} \times 1.00 \times 10^{-4} \text{ m}^3$
$= 0.092 \text{ kg}$

$\therefore$ The mass of water displaced $= 0.092 \text{ kg}$

$\therefore$ Volume of water displaced $= \dfrac{0.092 \text{ kg}}{1000 \text{ kg m}^{-3}} = 92 \text{ cm}^3$

(ii) When the ice melts, the melt water has the same density as the water in the can so occupies a volume of 92 cm³. Hence the water level stays the same. The principle is the same for a floating iceberg or ice-shelf. Hence melting sea ice hardly changes the sea level.

(c) The ocean surface is darker than ice, as is the rock exposed by retreating glaciers. Hence the fraction of the Sun's radiation absorbed by the Earth increases, raising the temperature further.

Q3 Some renewable sources, e.g wind, are intermittent. Solar energy has a reliable daily and seasonal variation but in Britain has a large intermittent element. Tidal power is renewable but almost totally dependable. It is not the intermittence which is the problem but the inability to store the generated energy, which could be overcome by using it produce a fuel, e.g. hydrogen by electrolysis of water.

Q4 (a) A fuel cell is a device in which the chemical energy in a fuel is used directly in a non-thermal process for the production of electrical energy.

(b) Fuel cells can be run from hydrogen, with oxygen from the atmosphere as the oxidising agent, so produce only water vapour as an exhaust gas. This is an advantage if the hydrogen is generated using electricity from a renewable source (or any CO_2 is sequestered). In conjunction with an electric motor, fuel cells are much more efficient than internal combustion engines.

Q5 (a) Uranium enrichment is increasing the percentage of ^{235}U in a sample of uranium from the natural level of 0.7%. It is necessary because the majority isotope, ^{238}U, is non-fissile but rather absorbs neutrons without fission. Fission reactors need about 3% of ^{235}U for normal operation.

(b) (i) The thorium nuclide must be $^{232}_{91}$Th.
$$^{232}_{91}\text{Th} + ^{1}_{0}\text{n} \rightarrow ^{233}_{91}\text{Th} \rightarrow ^{233}_{92}\text{U} + ^{0}_{-1}\text{e} + ^{0}_{0}\overline{\nu}_e$$

(ii) A ^{238}U nucleus absorbs a neutron, producing ^{239}U. This decays in two stages by β^- emission producing ^{239}Np followed by ^{239}Pu. The reactions are:
$$^{238}_{92}\text{U} + ^{1}_{0}\text{n} \rightarrow ^{239}_{92}\text{U}$$
$$^{239}_{92}\text{U} \rightarrow ^{239}_{93}\text{Np} + ^{0}_{-1}\text{e} + ^{0}_{0}\overline{\nu}_e$$
$$^{239}_{93}\text{Np} \rightarrow ^{239}_{93}\text{Pu} + ^{0}_{-1}\text{e} + ^{0}_{0}\overline{\nu}_e$$

Q6 (a)

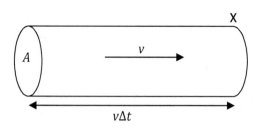

The mass of air passing the cross-section at **X** in time $\Delta t = Av\Delta t \times \rho$

∴ Kinetic energy of this air $= \frac{1}{2}(A\rho v\Delta t)v^2 = \frac{1}{2}A\rho v^3\Delta t$

∴ Dividing by Δt, the KE/second $= \frac{1}{2}A\rho v^3$

(b) Efficiency $= \dfrac{P_{OUT}}{P_{IN}} \times 100\%$

$$= \frac{7.9 \times 10^6 \text{ W}}{\frac{1}{2}\pi(80\,\text{m})^2 \times 1.25 \text{ kg m}^{-3} \times (12 \text{ m s}^{-1})^3} \times 100\%$$

$$= 36\%$$

Q7 (a) $\dfrac{\Delta Q}{\Delta t}$ = the heat flow per unit time (i.e. the power transfer).

Unit: W

$\dfrac{\Delta \theta}{\Delta x}$ = the temperature gradient, i.e. the temperature difference per unit length:

Unit: K m^{-1} or °C^{-1} m^{-1}

(b) $\left[\dfrac{\Delta Q}{\Delta t}\right] = [A][K]\left[\dfrac{\Delta \theta}{\Delta x}\right]$,

∴ W $= \text{m}^2 [K] \text{ K m}^{-1}$

∴ $[K] = \dfrac{\text{W}}{\text{m}^2 \text{ K m}^{-1}} = \text{W m}^{-1} \text{ K}^{-1}$

(c) Heat flows against the temperature gradient, i.e. from a high to a low temperature.

(d) $\Delta \theta = 25°\text{C} = 25 \text{ K}$

$$\frac{\Delta Q}{\Delta t} = 0.5 \text{ m}^2 \times 0.14 \text{ W m}^{-1} \text{ K}^{-1} \times \left(\frac{25 \text{ K}}{12.0 \times 10^{-3} \text{ m}}\right)$$

$$= 146 \text{ W} = 8750 \text{ J / minute}$$

Q8 (a) The intensity of the solar radiation at the ground is much less than 1361 W m^{-2} because of absorption and scattering by the atmosphere.

(b) At 15 V, in units of A and W m^{-2}: $\dfrac{9.2}{400} = 0.023$; $\dfrac{14.2}{600} = 0.024$; $\dfrac{19.2}{800} = 0.024$; $\dfrac{23.8}{1000} = 0.024$.

Hence proportional, within the tolerance of graphing.

(c) $P = VI$. At **X** $V = 0$, ∴ $P = 0$

At **Y** $I = 0$, ∴ $P = 0$.

(d)

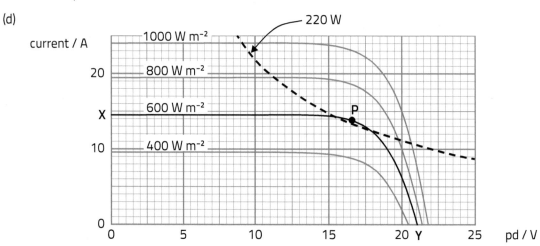

The broken line on the grid is for 220 W. The point **P** (15.5 V, 11.9 A) on the 600 W m^{-2} characteristic is slightly above the 220 W line, so the manufacturer's claim is correct.

The power at **P** is 16.5 V × 11.9 A = 229 W.

Q9 (a) (i) The kinetic energy of the water issuing per second $= \frac{1}{2}A\rho v^3$

But $v = \dfrac{\text{volume per second}}{\text{cross-sectional area}} = \dfrac{2.75 \text{ m}^3 \text{ s}^{-1}}{\pi \times (0.50 \text{ m})^2} = 3.50 \text{ m s}^{-1}$

$\therefore$ KE gain per second $= \frac{1}{2}\pi(0.5 \text{ m})^2 \times 1000 \text{ kg m}^{-3} \times (3.50 \text{ m s}^{-1})^3$

$= 16\,850$ W

(ii) Mass flow $= 2750 \text{ kg s}^{-1}$

$\therefore$ Loss in GPE per second $= 2750 \text{ kg s}^{-1} \times 9.81 \text{ N kg}^{-1} \times 6.0 \text{ m}$

$= 162\,000$ W

(b) Available power $= 162$ kW $- 17$ kW

$= 145$ kW

$\therefore$ Power output $= 80\% \times 145$ kW $= 116$ kW

(c) KE gained per second $= 16.85$ kW $\times (1.10)^3 = 22.43$ kW

GPE loss per second $= 162$ kW $\times 1.10 = 178$ kW

$\therefore$ Power output $= 80\% \times (178 - 22)$ kW $= 125$ kW

110% of 116 kW $= 128$ kW,

$\therefore$ output a bit less than 10% more

Original efficiency $= \dfrac{116 \text{ kW}}{162 \text{ kW}} \times 100 = 72\%$ (2 sf)

With 10% more flow: Efficiency $= \dfrac{125 \text{ kW}}{178 \text{ kW}} \times 100\% = 70\%$ (2 sf)

$\therefore$ Overall efficiency is lower but less than 10% lower.

Q10 The first stage is a weak interaction, shown by the emission of the neutrino, which is therefore much less likely to occur in any suitable collision than the second stage which is [a strong interaction followed by] an electromagnetic interaction.

Q11 (a) The rate of heat loss through each m² of the wall is 0.18 W for every °C difference in temperature between the inside and the outside.

(b) Rate of heat loss through wall $= (7.0 \times 2.2 - 3.0) \text{ m}^2 \times 0.18 \text{ W m}^{-2} \text{ K}^{-1}$

$= 2.23 \text{ W K}^{-1}$

Rate of heat loss through double-glazed units $= 4.5 \text{ W K}^{-1}$

Rate of heat loss through triple-glazed units $= 2.4 \text{ W K}^{-1}$

$\therefore$ % reduction $= \dfrac{4.5 - 2.4}{4.5 + 2.2} \times 100 = 31\%$

(c) The mean outdoor temperature is much lower in Norway, so the heat loss is much greater. Hence the payback time of installing triple glazing is much shorter.

Q12 (a) Rate of heat loss $= 8.0 \text{ m}^2 \times 0.62 \text{ W m}^{-1} \text{ K}^{-1} \times \dfrac{0.35 \text{ K}}{0.10 \text{ m}} = 17$ W

(b) Temperature difference across insulation $= \dfrac{0.62}{0.039} \times 0.35 °C = 5.56 °C$

$\therefore$ Total temperature difference $= 0.35 + 5.56 + 0.35 = 6.3 °C$ (2 sf)

(c) (i) $17 \text{ W} = 8.0 \text{ m}^2 \times (22 - 8) °C \times U$

$\therefore U = \dfrac{17 \text{ W}}{8.0 \text{ m}^2 \times 16 °C} = 0.13 \text{ W m}^{-2} \text{ K}^{-1}$

(ii)

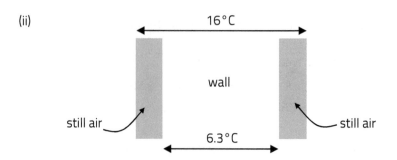

On either side of the wall there is a layer of still air which has a very low thermal conductivity and hence a large temperature difference across it.

(iii) The layer of still air on the outer face of the wall would be blown away so the temperature of the outer face would be much closer to the ambient air temperature.

Q13 (a) Peak emission $\lambda \sim 0.5$ μm.

Using Wien's law: $= T = \dfrac{W}{\lambda_{peak}} = \dfrac{2.90 \times 10^{-3} \text{ m K}}{0.5 \times 10^{-6} \text{ m}} = 5800$ K $= 6000$ K (1 sf)

(b) $\lambda_{peak} = \dfrac{W}{T} = \dfrac{2.90 \times 10^{-3} \text{ m K}}{290 \text{ K}} = 1.0 \times 10^{-5}$ m $= 10$ μm

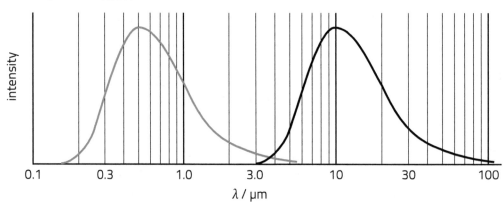

(c) Most of the solar radiation is between 0.4 and 0.9 μm, so reaches the surface of the Earth warming it up. The Earth's emitted radiation between 3 and 10 μm is partly absorbed by the atmosphere and re-emitted, 50% of which is downwards and warms the Earth further. The greater the concentration of the 'greenhouse gases', the greater the fraction of the Earth's radiation is absorbed and the higher the equilibrium temperature established – this is global warming.

Q14 (a) Height of tank = $\dfrac{\text{volume}}{\text{csa}} = \dfrac{400 \times 10^{-3} \text{ m}}{\pi \times (0.30 \text{ m})^2} = 1.415$ m

Area = area of top + area of sides
$= \pi \times (0.30 \text{ m})^2 + 2\pi \times 0.30 \text{ m} \times 1.415 \text{ m}$
$= 2.95 \text{ m}^2$

(b) The thermal conductivity of the steel is 1800× that of the PU and the thickness of the steel is much less (the surface areas of the steel and the PU are virtually the same).

(c) (i) $\dfrac{\Delta Q}{\Delta t} = 2.95 \text{ m}^2 \times 0.025 \text{ W m}^{-1} \text{K}^{-1} \times \dfrac{45°\text{C}}{0.025 \text{ m}} = 133$ W

(ii) $133 \text{ W} = 2.95 \text{ m}^2 \times 45 \text{ W m}^{-1} \text{K}^{-1} \times \dfrac{\Delta\theta}{0.003 \text{ m}}$

$\rightarrow \Delta\theta = 3 \times 10^{-3}$ °C ∴ justified.

[Alternatively: $45°\text{C} \times \dfrac{0.025}{45} \times \dfrac{0.3}{2.5} = 0.003°\text{C}$, ∴ justified.]

(d) The heat transfer from the thermal store, albeit small, raises the temperature of the cupboard, thus reducing the temperature gradient and hence the power loss.

Unit 3 Practice paper

1. (a) (i) The acceleration of a body is its rate of change of velocity.

   (ii) A car travelling at constant speed on a level, straight road. [There are many possible answers.]

   (b) (i) $\omega = \dfrac{2\pi \text{ rad}}{24 \text{ s}} = 0.26$ rad s^{-1}

   (ii) $v = rw = 60$ m $\times$ 0.262 rad s^{-1} = 16 m s^{-1}

   (iii) $a = rw^2 = 60$ m $\times$ (0.262 rad s^{-1})^2 = 4.1 m s^{-2}

   (c)

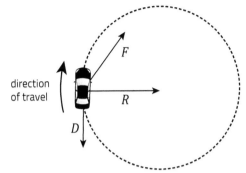

2. (a) (i) $2T = 3.15$ s $- 0.35$ s so $T = 1.40$ s

   $T = 2\pi\sqrt{\dfrac{m}{k}} \therefore k = \dfrac{4\pi^2 m}{T^2} = \dfrac{4\pi^2 \times 0.20 \text{ kg}}{(1.40 \text{ s})^2} = 4.0$ N m^{-1}

   (ii) $E_{k\,max} = \tfrac{1}{2}m(A\omega)^2 = \tfrac{1}{2}0.20$ kg $\times \left(0.080\text{m} \times \dfrac{2\pi}{1.40 \text{ s}}\right)^2 = 0.013$ J

   (iii) **Correct features:** E_k always positive. Maxima separated from minima by correct time interval (0.35 s).
   Incorrect features: Minima shouldn't be pointy; graph should be a sinusoid of period 0.7 s, shifted upwards. Maxima are shown at the times when the minima should be, and vice versa.

   (b) (i) Clamp a ruler vertically, close to path of sphere as it oscillates. Note reading of bottom of sphere against ruler at lowest point, viewing from same level to minimise parallax error. Subtract equilibrium reading of bottom of sphere. Repeat (releasing sphere pulled down by 0.080 m at $t = 0$) two or three times and take mean value of A.

   (ii) There are lots of methods but they all start from drawing the best-fit curve and noting that $A_0 = 60.0$ mm.
   Then, among these methods are (from the curve drawn):
   Method 1
   A (20 s) = 15.6 mm, $\therefore$ 15.6 = 60.0$e^{-20/\tau}$, $\therefore e^{-20/\tau} = 0.26$
   Taking logs $\rightarrow -\dfrac{20 \text{ s}}{\tau} = \ln 0.26 = -1.347$
   $\therefore \tau = 14.8$ s
   Method 2
   60 mm down to 15 mm is 2 half-lives = 20.6 s
   $\therefore$ Half-life = 10.3 s
   $\therefore \tau = \dfrac{10.3 \text{ s}}{\ln 2} = 14.9$ s
   Method 3
   τ is the time to drop to 1/e of the original value = 0.368
   0.368 $\times$ 60.0 mm = 22.1 mm
   $\therefore$ (from the graph) $\tau = 15.0$ s
   Note that an examiner will use your graph to work out a value for τ and mark your answer accordingly.

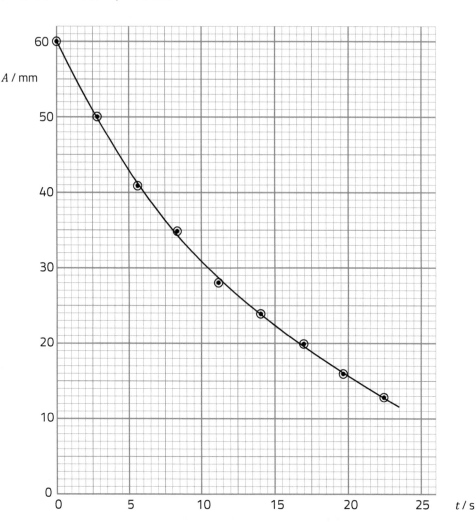

(iii) Plot ln A against t and draw a best fit straight line. It has a negative gradient = $-1/\tau$.
Hence $\tau = -1/$gradient.

3   (a) Heat flows from air in warmer room, through balloon skin into air in balloon. This takes place by molecules in room passing their greater mean KE on to molecules in skin and thence to air in balloon, in collisions between molecules. Thus mean KE of air molecules in balloon increases and these molecules hit inside of balloon skin more frequently and with greater change of momentum than previously, so the internal pressure increases. The balloon skin stretches (by rotation of bonds in the rubber molecules) so that the pressure inside is balanced by the combined pressure of balloon skin and outside air.

(b) (i) Density of helium, $\rho = \dfrac{6.0 \times 0.0040 \text{ kg}}{0.025 \text{ m}^3} = 0.96 \text{ kg m}^{-3}$

$p = \frac{1}{3}\rho \overline{c^2}$, $\therefore c_{rms} = \sqrt{\dfrac{3p}{\rho}} = \sqrt{\dfrac{3 \times 6.0 \times 10^5 \text{ Pa}}{0.96 \text{ kg m}^{-3}}} = 1.37 \text{ km s}^{-1}$

(ii) Temperature, T, of gas in cylinder = $\dfrac{pV}{nR} = \dfrac{6.0 \times 10^5 \text{ Pa} \times 0.025 \text{ m}^3}{6.0 \text{ mol} \times 8.31 \text{ Jmol}^{-1} \text{ K}^{-1}} = 300 \text{ K}$ (2 sf)

Gas temperature decreases when cylinder moved to store, so rms speed of molecules decreases.

4. (a) Q is the energy flowing into the system because of a temperature difference between the system and its surroundings.

(b) (i) Applying $pV = nRT$ at point A,

$n = \dfrac{pV}{RT} = \dfrac{9.0 \times 10^4 \text{ Pa} \times 4.0 \times 10^{-3} \text{ m}^3}{8.31 \text{ J mol}^{-1} \text{ K}^{-1} \times 280 \text{ K}} = 0.155 \text{ mol}$

(ii) Since the pressure is constant, $\dfrac{T_B}{T_A} = \dfrac{V_B}{V_A} = \dfrac{6.5}{4.0} = 1.625$

So $T_B = 1.625 \times 280 \text{ K} = 455 \text{ K}$. So $T_B - T_A = 455 \text{ K} - 280 \text{ K} = 175 \text{ K}$

(iii) $U = \frac{3}{2}nRT = \frac{3}{2}pV$, so at constant pressure,

$\Delta U = \frac{3}{2}p\Delta V = \frac{3}{2}9.0 \times 10^4\,\text{Pa} \times 2.5 \times 10^{-3}\,\text{m}^3 = 338\,\text{J}$

$W = p\Delta V = 225\,\text{J}$

$\therefore Q = \Delta U + W = 338\,\text{J} + 225\,\text{J} = 563\,\text{J}$

(iv) Since the initial and final states are the same as in (iii), ΔU is the same. But the gas does more work, as the area under the graph line is greater, so more heat must flow in.

5. (a) A is the number of nuclei decaying per unit time.

(b) When $t = T_{1/2}$, $A = \frac{1}{2}A_0$.

$\therefore \frac{1}{2}A_0 = A_0 e^{-\lambda/T_{1/2}}$

$\therefore e^{-\lambda/T_{1/2}} = \frac{1}{2}$

$\therefore -\lambda/T_{1/2} = \ln\frac{1}{2} = -\ln 2$

$\therefore \lambda = \frac{\ln 2}{T_{1/2}}$

(c) (i) $^{32}_{15}P\ P \rightarrow\ ^{32}_{16}S\ +\ ^{0}_{-1}e\ +\ ^{0}_{0}\bar{v}$

(ii) (I) $A = \lambda N$, $N = \frac{A}{\lambda} = \frac{AT_{1/2}}{\ln 2} = \frac{240 \times 10^9\,\text{Bq} \times (14.2 \times 24 \times 3600)\,\text{s}}{\ln 2}$, $= 4.25 \times 10^{17}$

Mass of $^{32}_{15}P$ atom $= 32\,\text{u} = 32 \times 1.66 \times 10^{-27}\,\text{kg}$

$\therefore$ Mass of sample $= 4.25 \times 10^{17} \times 32 \times 1.66 \times 10^{-27}\,\text{kg} = 2.3 \times 10^{-8}\,\text{kg}$ (23 µg)

(II) From 240 Bq to 180 Bq is a loss of a quarter of the initial activity.
But the rate of loss is greater for the first quarter than the second.
So the drop from 240 Bq to 180 Bq takes less than half the half-life and Rhian is correct.

6. (a) (i) Mass defect of $^{7}_{3}Li = (3 \times 1.00728 + 4 \times 1.00866 - 7.01435)\,\text{u} = 0.04213\,\text{u}$
So BE $= 0.04213 \times 931\,\text{MeV} = 39.22\,\text{MeV}$,

and BE/N $= \frac{39.22\,\text{MeV}}{7} = 5.60\,\text{MeV}$

(ii) (I)

(II) In both fission and fusion, the number of nucleons stays the same but binding energy per nucleon increases (see diagram), so the mass energy in the nuclei decreases. Since energy is conserved, the mass energy lost is released as kinetic energy.

(b) (i) 0.800 MV

(ii) Total KE of $^{4}_{2}He$ = KE of p + mass energy of p + mass energy of $^{7}_{3}Li$
$- 2 \times$ mass energy of $^{4}_{2}He$

$\therefore$ Total KE of $^{4}_{2}He = 0.800\,\text{MeV} + (1.00728 + 7.01435 - 2 \times 4.00151) \times 931\,\text{MeV}$
$= 18.1\,\text{MeV}$

(iii) To conserve momentum, the $^{4}_{2}He$ released in the direction in which the proton was travelling must have more momentum than the $^{4}_{2}He$ released in the opposite direction. But the masses are equal, so the first $^{4}_{2}He$ must be moving faster and have more KE than the second.

Unit 4 Practice paper

SECTION A

1. (a)

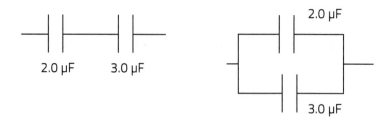

$$C = \frac{2.0\,\mu F \times 3.0\,\mu F}{2.0\,\mu F + 3.0\,\mu F} = 1.2\,\mu F \qquad C = 2.0\,\mu F + 3.0\,\mu F = 5\,\mu F$$

(b) Loss of energy by 10 F capacitor $= \frac{1}{2}CV_1^2 - \frac{1}{2}CV_2^2$

$$= \frac{1}{2} \times 10(6^2 - 4.5^2)\,J$$

$$= 79\,J$$

Energy needed for 2 minute of brushing = 0.75 W × 120 s = 90 J
∴ 10 F is too small.

(c) (i) Dividing through the given equation by C: $\frac{Q}{C} = \frac{Q_0}{C}e^{-t/CR}$, ∴ $V = V_0 e^{-t/CR}$
Taking logs of the [numerical] quantities on both sides:
$\ln(V/V) = \ln(V_0/V) + \ln(e^{-t/CR})$
∴ $\ln(V/V) = \ln(V_0/V) - \frac{1}{CR}t$

(ii) Extreme values of pd are 5.43 V and 5.88 V (and these are roughly equally spaced about the mean value of 5.64 V).
ln 5.43 = 1.69; ln 5.88 = 1.77 so error bar is correctly plotted.

(iii)

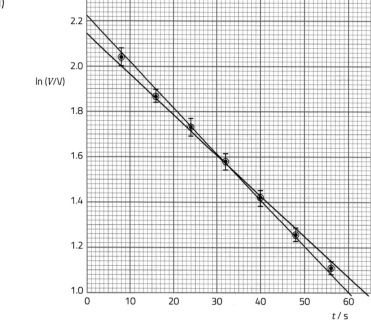

$$\text{gradient} = -\frac{1}{CR}$$

$$\therefore C = -\frac{1}{\text{gradient} \times R}$$

$$\text{Min gradient} = \frac{1.00 - 2.22}{60.0\,s} \; ; \text{Max gradient} = \frac{1.00 - 2.14}{63.8\,s}$$

$$\therefore \text{Min } C = \frac{60.0\,s}{1.22 \times 68\,000\,\Omega} = 723\,\mu F; \text{Max } C = \frac{63.8\,s}{1.14 \times 68\,000\,\Omega} = 823\,\mu F$$

So $C = (770 \pm 50)\,\mu F$ (2 sf)

2. (a) $E = \dfrac{\text{force on (test) charge}}{\text{(test) charge}}$

(b) $E = \dfrac{Q}{4\pi\varepsilon_0 r^2}$,

$\therefore Q = 4\pi\varepsilon_0 r^2 E = 4\pi \times 8.85 \times 10^{-12} \times 0.10^2 \times 2.0 \text{ C}$
$= 2.22 \text{ pC}$

(c)
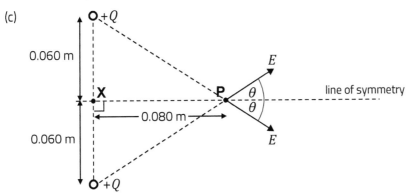

$r = \sqrt{0.060^2 + 0.080^2}$ m = 0.10 m. So E (see diagram) is 2.0 N C^{-1}

$\therefore$ Resultant field strength at **P** $= 2E \cos\theta = 2 \times 2.0 \times \dfrac{0.080}{0.100}$
$= 3.2$ N C^{-1} to the right

(d) KE gained by ion = Electrical PE lost. $\therefore \frac{1}{2}mv^2 = 2 \times \dfrac{Qq}{4\pi\varepsilon_0 \times 0.060 \text{ m}}$

$\therefore v = \sqrt{\dfrac{4Qq}{4\pi\varepsilon_0 m \times 0.060 \text{ m}}} = \sqrt{9.0 \times 10^9 \dfrac{4 \times 2.22 \times 10^{-12} \times 1.60 \times 10^{-19}}{1.05 \times 10^{-25} \times 0.060}}$ m s^{-1}
$= 1.4$ km s^{-1}

(e) When the ion sets out from X, $\cos\theta$ increases from zero, and so does the *resultant* field strength and the ion's acceleration. Elodie has missed this. But as the ion travels on, the (inverse square law) decrease in field strength due to increasing distance from the charges (Q) becomes the important factor, and the ion's acceleration does decrease (asymptotically to zero).

3. (a) (i) Period = 3.2 year = $3.2 \times 3.16 \times 10^7$ s = 1.01×10^8 s
orbital speed = 120 m s^{-1}
So star's orbital radius = $\dfrac{120 \text{ m s}^{-1} \times 1.01 \times 10^8 \text{ s}}{2\pi}$ = 1.93×10^9 m

(ii) Assuming the planet's mass is negligible
$\therefore T = 2\pi\sqrt{\dfrac{d^3}{GM_s}}$, $\therefore d^3 = \left(\dfrac{1.01 \times 10^8}{2\pi}\right)^2 \times 6.67 \times 10^{-11} \times 4.67 \times 10^{30}$ m^3
$\therefore d = 4.32 \times 10^{11}$ m

(iii) $m_p r_p = m_s r_s$
But $m_s \gg m_p$ so $r_s \ll r_p$ so to a good approximation, $r_p = d$
$\therefore m_p = m_s \dfrac{r_s}{d} = 4.67 \times 10^{30}$ kg $\times \dfrac{1.93 \times 10^9 \text{ m}}{4.32 \times 10^{11} \text{ m}} = 2.1 \times 10^{28}$ kg

(b) It's natural to wonder whether there is intelligent life on other planets, and there would be popular support for projects which could lead eventually to its discovery.

But because of the likely long distance away of any Earth-like exoplanets, and the finite speed of light, two-way communication is unlikely ever to be practicable.

Other equally, or more, worthwhile and more fundamental projects might suffer.

4. (a) $H_0 = \dfrac{\text{recession speed of a galaxy}}{\text{distance of galaxy from us}}$

(b) (i) $\dfrac{v}{c} = \dfrac{\Delta\lambda}{\lambda}$ and $H_0 = \dfrac{v}{D}$

so $D = \dfrac{c}{H_0}\dfrac{\Delta\lambda}{\lambda} = \dfrac{3.00 \times 10^8 \text{ m s}^{-1}}{2.20 \times 10^{-18} \text{ s}^{-1}} \times \dfrac{694 - 656}{656} = 7.9 \times 10^{24}$ m

(ii) Age of universe $= \dfrac{1}{H_0} = \dfrac{1}{2.20 \times 10^{-18} \text{ s}^{-1}} = 4.54 \times 10^{17} \text{ s} \, [14 \times 10^9 \text{ year}]$

The universe has not been expanding at a constant rate.

(c) $\left[\dfrac{3H_0^2}{8\pi G} \right] = \dfrac{(\text{s}^{-1})^2}{\text{N kg}^{-2}\text{m}^2} = \dfrac{\text{s}^{-2}}{(\text{kg m s}^{-2}) \text{ kg}^{-2} \text{ m}^2} = \text{kg m}^{-3} = [\rho_c]$

$\therefore$ Homogeneous

5. (a) (i) Current, from the positive of the battery, will be in the XY direction.
According to Fleming's left hand motor rule the force on the wire will be upwards, so the battery doesn't need to be reversed.

(ii) For equilibrium in original position with switch closed, $BI\ell = mg$

$\therefore B = \dfrac{mg}{I\ell} = \dfrac{4.50 \times 10^{-3} \times 9.81}{4.20 \times 0.15} \text{ T} = 70 \text{ mT}$

(b) As the rod moves upwards it cuts magnetic flux, so an emf is induced in it. If the switch is closed, the rod forms part of a complete conducting circuit, WXYZ. So there is a current in the rod, which is from right to left at a (right) angle to the field direction. Hence the rod experiences a ('motor effect') force downwards.

6. (a) The moving charge carriers constitute a current from left to right, so are deflected upwards by the magnetic field. So the top of the wafer is given a positive charge and the bottom face, depleted of charge carriers, a negative charge. Hence, there is a downward electric field, E.

Soon carriers are no longer deflected, because we have a steady state, when

Electric force on charge carrier $= -$ magnetic force on charge carrier

So the magnitudes of these forces are equal: i.e. $Eq = Bqv$

But $E = \dfrac{\text{pd between } \mathbf{X} \text{ and } \mathbf{Y}}{\text{distance between } \mathbf{X} \text{ and } \mathbf{Y}} = \dfrac{V_H}{b}$

$\therefore \dfrac{V_H}{b} = Bv$ that is $V_H = bBv$

(b) We know that $I = nAvq$ in which, in this case, $A = ab$, $\therefore v = \dfrac{I}{nabq}$.

We see that, if a is smaller or if n is smaller, then v will be larger. If either is the case V_H will be larger. [The effects of changing b cancel.]

(c) The measured Hall pd when there is no current through the wire is clearly a zero error. Subtracting it from the first three readings reduces these to 65, 42, 32.

Because $B = \dfrac{\mu_0 I}{2\pi r}$, in which $\dfrac{\mu_0 I}{2\pi}$ is constant, Br and therefore rV_H should be constant.

Checking $(r/\text{mm}) \times (V_H/\text{V})$: $40 \times 65 = 2600$; $60 \times 42 = 2520$; $80 \times 32 = 2560$

These results are consistent with the expected inverse law, bearing in mind that the experimental data are only to 2 significant figures.

7. (a)

(b) (i) $B = \mu_0 n I$,

$\therefore \dfrac{\Delta B}{\Delta t} = \mu_0 n \dfrac{\Delta I}{\Delta t} = 4\pi \times 10^{-7} \times \dfrac{400}{0.80} \times 0.60 \text{ T s}^{-1} = 3.77 \times 10^{-4} \text{ T s}^{-1}$

(ii) $|\text{emf}| = \dfrac{\Delta \Phi}{\Delta t} = N\pi r^2 \dfrac{\Delta B}{\Delta t} = 250 \times \pi \times 0.015^2 \times 3.77 \times 10^{-4} \text{ V} = 67 \text{ μV}$

Option A: Alternating currents

8. (a)

Note that the phase is arbitrary.

(b) (i) Total impedance $Z = \sqrt{R^2 + X^2} = \sqrt{30^2 + 40^2} = 50\ \Omega$

∴ Current, $I = \dfrac{V}{Z} = \dfrac{12\ V}{50\ \Omega} = 0.24\ A$

∴ $V_R = 0.24\ \Omega \times 30\ \Omega = 7.2\ V$ and $V_C = 0.24\ \Omega \times 40\ \Omega = 9.6\ V$

(ii) The two voltages are [90°] out of phase with each other $\left[\text{so } V_{tot} = \sqrt{V_R^2 + V_C^2} \right]$.

(iii) The total reactance is $|X_L - X_C|$, which is less than X_C as long as $X_L < 2X_C$. Thus the total impedance will be less, the current will be more and so the pd across the resistor will be more. But if $X_L > 2X_C$, the total impedance will be more and the pd across the resistor will be less.

(c) (i) Nigel has increased [doubled] the Y gain. He has also adjusted the Y position, so the top of the waves are on a grid line and the X position so the position of the bottom of the waves is easily read from the scale. These all reduce the uncertainty in the peak voltage and hence in the current.

(ii) (I) 2 periods = 6.2 Divisions = 124 μs, ∴ Period = 62 μs

∴ Frequency = $\dfrac{1}{62\ \text{μs}}$ = 16 kHz (2 sf)

(II) Amplitude = $\dfrac{1}{2} \times 5.5$ divisions = 27.5 mV

∴ Peak current = $\dfrac{27.5\ \text{mV}}{12\ \text{k}\Omega}$ = 2.29 μA

∴ Rms current = 1.6 μA (2 sf)

(iii) If the time base were increased to 10 μs div^{-1} there would only be one whole cycle on the screen with a cycle length of 7.6 divisions, so it is no more accurate.

Option B: Medical physics

9.   (a) (i)   Time lag = 32 − 11 = 21 ms

Distance = speed × time = 1620 m s^{-1} × 21 × 10^{-6} s = 34.02 mm

Since this is a reflection, the actual distance is 17 mm

(ii)   Ratio of intensities = $\dfrac{(1.74 - 1.53)^2}{(1.74 + 1.53)^2}$ = 0.004 12, i.e. 0.412 %

(iii)   The acoustic impedance of tissue is a million times greater than the acoustic impedance of air. This means that nearly all ultrasound is reflected at air-tissue boundaries. Bubbles between the transducer and the skin will cause 100% reflection and ruin the scan. This is avoided by using a coupling gel (which is designed to have a similar acoustic impedance to tissue).

(b) (i)   The minimum wavelength = 30 pm from the graph.

$$eV = \frac{hc}{\lambda} \rightarrow V = \frac{hc}{e\lambda} = \frac{6.63 \times 10^{-34} \text{ J s} \times 3 \times 10^8 \text{ m s}^{-1}}{1.60 \times 10^{-19} \text{ C} \times 30 \times 10^{-12} \text{ m}} = 41 \text{ kV}$$

(ii)   The line spectrum is caused by an inner electron of the target being knocked out by one of the 41 keV electrons. Another electron can then drop down from a higher energy level to take its place. [Both energy levels are narrow so that these will be very specific wavelengths.]

(c) (i)   Larmor frequency = 42.6 MHz T^{-1} × B = 42.6 MHz T^{-1} × 1.50 T = 63.9 MHz

(ii)   So the Larmor frequency varies along the body of the patient. A certain Larmor frequency will scan a certain 'slice' of the patient – producing a 2D image of that slice. [As the Larmor frequency varies, these 2D slices will build up a 3D image of the patient.]

(iii)   Different tissues have different concentrations of hydrogen nuclei [due to different water concentrations] which produce different relaxation times. [This is the time it takes for the hydrogen nuclei to flip back to their original alignment with the magnetic field.] The MRI scanner detects these differences.

(d)   Advantage of an MRI scanner is that no ionising radiation is involved. A disadvantage of an MRI scanner is that it cannot be used on patients with metal implants or pace-makers (although MRI friendly pace-makers have been available since 2011).

Option C: The physics of sports

10.   (a) The paddle boarder is stable as long as her centre of gravity is above the 'base'. The CoG is high compared to the width of the base so a small movement will cause the paddle board to tilt (reducing the width of the base further) and taking the CoG outside the base.

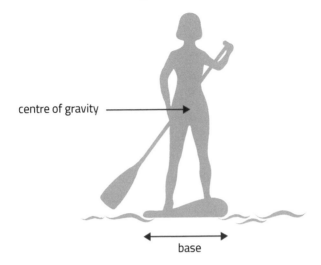

centre of gravity

base

(b) (i)   The air is deflected upwards by the wings, gaining a vertical momentum. Force is rate of change of momentum (Newton's second law), so the car exerts an upward force on the air, which exerts an equal and opposite (i.e. downward) force on the car, by Newton's third law.

The car is not accelerating vertically, so there is no resultant vertical component of force. The normal contact force is equal to the sum of the car's weight and the downward force produced by the wings, so is larger than if the wings were not there.

   (ii) Increasing the normal contact force increases the maximum grip, which gives a greater maximum acceleration, hence greater mean speeds in straight sections. When negotiating bends, the grip provides the centripetal force, so a greater grip allows higher cornering speeds. Greater mean speed gives a smaller lap time.

(c) (i)   Horizontal component of velocity = 6.50 cos 65° m s^{-1}
No horizontal forces on the ball, so the horizontal component of velocity is constant.

$$\therefore \text{Time taken} = \frac{x}{t} = \frac{3.0\,\text{m}}{6.50\cos 65° \text{m s}^{-1}} = 1.09(2)\,\text{s}$$

   (ii) Height after time $t = 2.50 + ut\sin\theta - \frac{1}{2}gt^2$

So after 1.135 s, h = 2.50 m + 6.5 m s^{-1} × 1.092 s cos 65° − 4.905 m s^{-2} × (1.092 s)^2
            = 3.08 m
This is virtually the same height as the loop, so the ball will bounce off the loop and not go through.

Option D: Energy and the environment

11. (a) (i) The density of water is much greater than the density of air (1000 kg m^{-3} compared to 1.2 kg m^{-3}, i.e. about 800×). The speed of water in tidal streams is comparable to the speed of wind.

$\therefore$ For the same power: $\frac{A_{water}}{A_{air}} \sim \frac{1}{800}$ $\therefore \frac{l_{water}}{l_{air}} \sim \frac{1}{\sqrt{800}} \sim \frac{1}{30}$, where l is the length of the turbine blades.

[Note that full marks could be obtained for a good qualitative answer to this question.]

(ii) (I) For a given volume flow rate, the narrower the channel the greater the speed of flow. The proposed site is at the narrowest point where the water has the greatest kinetic energy.

(II) Whereas wind is very variable, the tides are have predictable heights and therefore the power generation is predictable. There will be four periods of greatest predictable flow per day.

(III) Around high and low tides, there will be no flow ('slack water') and hence no generation. Also the magnitude of the flow will vary throughout the day necessitating energy storage or backup.

(IV) $P_{out} = 0.35 \times \frac{1}{2}\pi l^2 \rho v^3$

$\therefore l^2 = \frac{2P_{out}}{0.35\pi\rho v^3} = \frac{2.0 \times 3.0 \times 10^6 \text{ W}}{0.35\pi \times 1020 \text{ kg m}^{-3} \times (8.0 \text{ m s}^{-1})^3}$

$\therefore l = 3.2$ m (2 sf)

(b) (i) Nuclear fission is the process whereby the nucleus of a fissile nuclide, e.g. ^{235}U, absorbs a slow (thermal) neutron and splits into two massive fragments and several neutrons, with the release of a large amount of energy.

Nuclear fusion is the process whereby two light nuclei collide at high enough energy to approach closely and combine to give a single heavier nucleus (often with the emission of a neutron) with the release of a large amount of energy.

(ii) The nuclei have to approach within 10^{-14} m so need to be at a high temperature ($\sim 10^8$ K), so need to be isolated from the container walls. The number density has to be high to ensure a large number of collisions for a long period of time.

(c) Let the temperature of the concrete/PU boundary be θ. The heat flow through the concrete and PU are equal, so $\frac{\Delta Q}{\Delta t} = A \times 0.92 \text{ W m}^{-1}\,°\text{C}^{-1} \times \frac{\theta - (-5\,°\text{C})}{3.5 \text{ cm}} = A \times 0.034 \text{ W m}^{-1}\,°\text{C}^{-1} \times \frac{15\,°\text{C} - \theta}{2.5 \text{ cm}}$

$\therefore 19.3 \, (\theta + 5\,°\text{C}) = 15\,°\text{C} - \theta$

$\therefore 20.3\theta = -81.5\,°\text{C}$

$\therefore \theta = -4.01\,°\text{C}$

$\therefore$ Heat flow through PU, $\frac{\Delta Q}{\Delta t} = 20 \text{ m}^2 \times 0.034 \text{ W m}^{-1}\,°\text{C}^{-1} \times \frac{19.01\,°\text{C}}{2.5 \times 10^{-2}\text{m}} = 520$ W (2 sf)

This is less than 1 kW, so the suggestion is valid.

Notes

Notes